Lecture Notes in Earth Sciences

Lecture Notes in Earth Sciences

Edited by Somdev Bhattacharji, Gerald M. Friedman,
Horst J. Neugebauer and Adolf Seilacher

6

Werner Ricken

Diagenetic Bedding

A Model for Marl-Limestone Alternations

Springer-Verlag
Berlin Heidelberg GmbH

Author

Dr. Werner Ricken
until October 1986:
University of Colorado, Department of Geological Sciences
Campus Box 250, Boulder, C0 80309, USA

after October 1986:
Universität Tübingen, Institut für Geologie und Paläontologie
Sigwartstr. 10, D-7400 Tübingen, FRG

ISBN 978-3-540-16494-4   ISBN 978-3-540-46554-6 (eBook)
DOI 10.1007/978-3-540-46554-6

© Springer-Verlag Berlin Heidelberg 1986
Originally published by Springer-Verlag Berlin Heidelberg New York in 1986

2132/3140-543210

PREFACE

The study of calcareous bedding rhythms has become an important field in Geology. Often these bedding rhythms are simply interpreted as representations of primary climatic cycles without showing the effects of any appreciable diagenetic overprinting. This study, however, deals predominantly with the diagenetic processes which are usually large and affect both the amplitude and rhythm of carbonate oscillations. The purpose of this textbook is twofold. First, it intends to provide a better understanding of the processes of diagenetic bedding. Secondly, this new approach allows one to quantify and to understand diagenesis in terms of mass exchanges. This is possible through the development of methods which combine chemical data with compaction measurements. These methods can be also used independent of the marl-limestone alternation problem.

This book is an updated translation of my dissertation which was partially funded by the "Sonderforschungsbereich Palökologie" at the Department of Geology and Paleontology (Tübingen, W. Germany). My first ideas about diagenetic bedding were formulated during core examinations of stylolitic marl-limestone alternations which were carried out several years ago with C. HEMLEBEN (Tübingen) and during discussions with W. EDER and M. WALTHER (both at Göttingen) spanning the years 1979 to 1981. After my return from Göttingen University to Tübingen, the major ideas for this study developed in response to cyclic sedimentation models. I gratefully thank G. EINSELE (Tübingen), who promoted the study in critical discussions and who always provided encouraging support. I also wish to thank C. HEMLEBEN, F. LIPPMANN, and A. SEILACHER (all Tübingen) for providing valuable ideas. The field trips with A.G. FISCHER (Los Angeles) and T. HERBERT (Princeton) in Italy, who showed me the Gubbio section and explained their view of a preferentially cyclic formation of marl-limestone alternations, were very stimulating. R. BATHURST (Liverpool), who independently developed a similar model about diagenetic bedding, gave a lot of support and helped where he could. I am grateful to my colleagues at the Boulder Department of Earth Sciences (Colorado), where I did a part of the translation work, who sustained me in many ways, especially R. DINER, D. EICHER, W. ELDER, P. HARRIES, E. KAUFFMAN, J. KIRKLAND, B. SAGEMAN, T. WALKER. The author would also like to acknowledge several colleagues for discussions, written communications, and field trips. These are T. BARCHERT (Erlangen), U. BAYER (Tübingen), P. COTILLON (Lyon), W. FRANKE (Göttingen), E. FLÜGEL (Erlangen), H. FÜCHTBAUER (Bochum), B. LANG (Erlangen), H.A. LOWENSTAM (Pasadena), H.P. LUTERBACHER (Tübingen), D. MEISCHNER (Göttingen), G. NAPOLEONE (Florence), J. NEUGEBAUER (Tübingen), U. ROSENFELD (Münster), W. SCHWARZACHER (Dublin), J. VEIZER (Ottawa), H.R. WANLESS (Miami), J. WIEDMANN (Tübingen), W. WILLE (Tübingen), and A. WETZEL (Tübingen). The exchange of ideas with my fellow Tübingen doctoral students was instructive, especially with D. RUPP (field trip to southern France), T. AIGNER, C. RUCH, as well as G. GEBHARD and W. RIEGRAF (who both helped with the identification of fossils).

M. HECKENBERGER, H. WINDER (both of Tübingen) and L. WITTOCK (Brussels) did some of the analytical work and helped to reduce the extensive number of raw samples. The acetic acid disintegration and the determination of minor elements were performed in the Geochemical Central Laboratory together with M. FETH and H. FRIEDRICHSEN. The sampling of interesting quarry walls was possible due to T. RATHGEBER (Ludwigsburg), who supplied me with a steel rope ladder and climbing rope. W. WETZEL (Tübingen) made reproductions of some of the figures. Several persons were very helpful with improving the English text and with correcting the proofs. These are P. HARRIES (Boulder), E. HERRMANN, L. HOBERT, and H. WINDER (all Tübingen). J. SAFFELL (Boulder) typed the text of the English edition. To all these, my grateful thanks.

Boulder, Colorado   Werner Ricken
March 1986

C O N T E N T S

SYMBOLS OF THE MOST COMMONLY USED PARAMETERS

a, b	Large (a) and small (b) axes of deformed, originally cylindrical burrows (perpendicular to the burrow tube).
C	Carbonate content (volume or weight%).
C_d	Carbonate content expressed as a percentage of the decompacted (primary) sediment volume (vol%).
C_n	Statistically neutral carbonate content at the boundary between dissolution and cementation zones (%).
C_w	Carbonate content at the weathering boundary between marl and limestone (%).
C_o	Carbonate content of the primary sediment (vol%).
C_{od}	Primary carbonate content expressed as a percentage of the decompacted (original) sediment volume (vol%).
Ca_s, Ca_{cc}	Amount of calcium in a solution (s) and in calcite (cc) which is precipitated from this solution.
D	Burrow deformation, expressed as the amount of reduction of the original thickness (%).
d	Density gradient in existing carbonate ooze, which is dependent on overburden, after HAMILTON (1976).
e	Quotient describing the relative enrichment of insoluble particles in a given volume of the marl beds relative to that of the adjacent limestone layers.
F	Factor calculating the diagenetic enhancement between original and postdiagenetic carbonate fluctuations.
h	Thickness of a compacted rock interval.
h^*	Thickness of a decompacted rock interval.
K	Compaction of primary sediment volume by a certain amount (vol%).
K_1	Mean compaction in the middle of the limestone layers in an alternation, defined as the amount of compaction at the onset of lithification. This is equivalent to the mechanical compaction (MK, vol%).
K_n	Compaction at the statistically neutral zone between the intervals of carbonate dissolution and cementation (vol%).
K_L, K_M	Compaction in the middle of adjacent limestone and marl layers (vol%).
K_{-Z}, K_Z	Mean compaction in the zones of carbonate dissolution and cementation within a given marl-limestone alternation (vol%).
k_{TE}^{cc}	Distribution coefficient describing the molar ratio of a trace element (TE) and calcium between the precipitating mineral phase (cc) and the solution.
MK	Mechanical compaction, equivalent to the degree of compaction at the onset of lithification K_1 (vol%).
n	Porosity, expressed as a percentage of the rock or sediment volume (vol%).
n_d	Porosity, expressed as a percentage of the decompacted (primary) sediment volume (vol%).
n_1	Porosity of the sediment at the onset of lithification (vol%).
n_o	Mean decompaction porosity (vol%) resulting from numerical decompaction of the entire section studied.
N	Amount of insoluble particles contained in a given (compacted) rock volume.

ΔN	Increase of insoluble particles contained in a given volume of rock after compaction.
N_L, N_M	Amount of insoluble particles contained in a given volume of rock in adjacent limestone (L) and marl beds (M).
N_o	Original amount of insoluble particles contained in an uncompacted sediment or rock volume.
NC	Noncarbonate fraction, expressed as a percentage of the total volume of solids (vol%).
NC_d	Noncarbonate fraction, expressed as a percentage of the decompacted (primary) sediment volume; also referred to as "absolute clay content" (vol%).
NC_{dr}	Noncarbonate fraction, expressed as a percentage of the decompacted (primary) sediment volume in dense or nearly dense rock with no or low porosity (vol%).
P,R	"Primary" carbonate fraction (P) within the cementation zone, relic carbonate fraction (R) within the dissolution zone.
ρ_h	Density of sediment (g/cm^3) with overburden (h).
ρ_m	Density of mineral grains (g/cm^3).
ρ_o	Density of the decompacted (primary) sediment (g/cm^3).
S	Amount of pore-free solids (vol%).
σ	Standard deviation.
t	Time since deposition.
TE	Concentration of a given minor element contained in the carbonate fraction (ppm).
ΔTE	Increase or decrease in concentration of a given minor element contained in the carbonate fraction due to chemical compaction (ppm).
TE_n	Concentration of a given minor element contained in the carbonate fraction at the neutral boundary between dissolution and cementation zones (ppm).
TE_P, TE_R	Concentration of a given minor element contained in the "primary" (P) and relic (R) carbonate fractions, respectively (ppm).
TE_s, TE_{cc}	Concentration of a given minor element contained in the solution (s) and in the precipitating calcite phase (cc), respectively (ppm).
TE_{-Z}, TE_Z	Concentration of a given minor element contained in the dissolved (-Z) and cemented (Z) carbonate fractions, respectively (ppm).
V_{-Z}, V_Z	Decompacted (primary) sediment volume for dissolution zones (V_{-Z}) and cementation zones (V_Z), respectively.
X	Degree of closure in the system of the dissolution zones to influx or outflux of a given minor element contained in the carbonate fraction (%). 100% closure is equivalent to a complete retainment of a given minor element concentration in the dissolution zone.
z	Cement number, the ratio between the normalized amounts of the dissolved or cemented carbonate fraction (Z_d) and the primary carbonate fraction (C_{od}).
-Z,Z	Dissolved carbonate fraction (-Z) and cemented carbonate fraction (Z); also used to describe the processes of dissolution and cementation.
$-Z_d, Z_d$	Amounts of dissolved ($-Z_d$) and cemented carbonate (Z_d), respectively, expressed as a percentage of the decompacted (primary) sediment volume (vol%); also referred to as the "absolute amounts" of dissolved and cemented carbonate (vol%).
$-Z_c, Z_c$	Relative amounts of dissolved ($-Z_c$) and cemented carbonate (Z_c), respectively, expressed as a percentage of the total carbonate content (vol or weight%).

1 INTRODUCTION

1.1 Concept of Diagenetic Bedding

Numerous contradictory explanations have been proposed for the widespread phenomenon of rhythmic bedding, particularly for marl-limestone alternations (EINSELE, 1982). In the past, efforts were made to distinguish such alternations from sequences which are generated by repeated depositional events (EINSELE & SEILACHER, 1982). Later, cyclic depositional processes were thought to cause rhythmic bedding in marl-limestone alternations. Such explanations are usually based on a model of rhythmic climatic oscillations, named Milankovitch cycles, which form by variations in the Earth's orbital parameters and have periodicities of 20, 40, 100, and 400Ka (e.g., GILBERT, 1895;

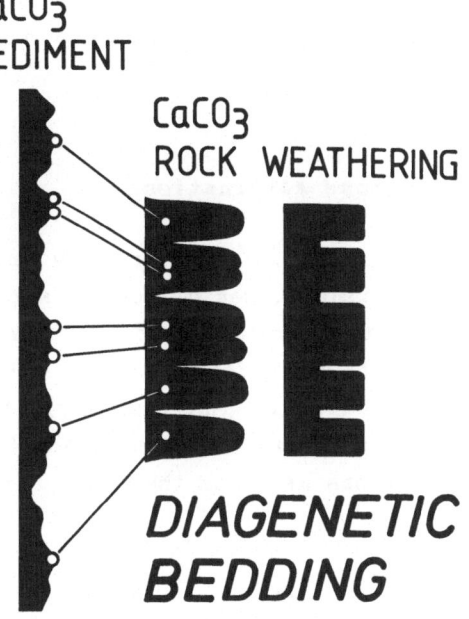

Fig. 1 Concept of "diagenetic bedding". Diagenetic transformation of primary sediment (left) into a rhythmic marl-limestone alternation (right).

MILANKOVITCH, 1930; FISCHER & ARTHUR, 1977; FISCHER, 1980; EINSELE, 1982; SCHWARZACHER & FISCHER, 1982; BERGER et al., 1984; COTILLON & RIO, 1984; FISCHER et al., 1985).

This study will attempt to explain the existence of marl-limestone alternations in terms of repeating zones of diagenetic carbonate dissolution and cementation. These processes considerably enhance slight primary carbonate variations and thereby generate rhythmic bedding (i.e., "diagenetic bedding" or "diagenetische Bankung"; Fig. 1). According to this theory, it is only of minor importance whether the original bedding was cyclic or stochastic, because the processes which produce diagenetic bedding always generate sequences with more pronounced rhythmicity than that found in the primary sediment. In principle, similar concepts have frequently been used to explain marl-limestone alternations (e.g., WEPFER, 1926; SUJKOWSKI, 1958; BARRET, 1964; HALLAM, 1964; HENNIGSMOEN, 1974; CAMPOS & HALLAM, 1979; TRURNIT & AMSTUTZ, 1979; EDER, 1982; WALTHER, 1982; BATHURST, 1984; GLUYAS, 1984; SIMPSON, 1985). However, these attempts have been more or less solely qualitative and are somewhat unconvincing. In the present study, therefore, new methods have been developed to quantify dissolution and cementation processes by using carbonate and minor element mass balance calculations. These calculations, which are discussed in section 2, are based on methods to evaluate rock compaction and on the mathematical relationship between compaction, carbonate content, and porosity.

1.2 Studied Marl-Limestone Alternations

In order to eliminate regional pecularities and to produce more generally applicable conclusions, ten different sections of micritic marl-limestone alternations with various degrees of diagenetic overprint were selected from different areas. These are from the South German and Lower Saxony Basins (Germany), the Vocontian Basin (French Maritime Alps), the Umbrian Apennines, and Sicily (Italy, Fig. 2). The studied alternations are from the Upper Jurassic, Cretaceous,

Fig. 2 Marl-limestone alternations studied. R=Rheine (Upper Cretaceous), N=Neuffen (Upper Jurassic), GS=Geisingen (Upper Jurassic), A=Angles (Lower Cretaceous), L=Logis du Pin (Lower Cretaceous, F=Fossombrone (Maastrichtian to Paleogene), G=Gubbio (Maastrichtian to Paleogene), PE=Porto Empedocle (Pliocene).

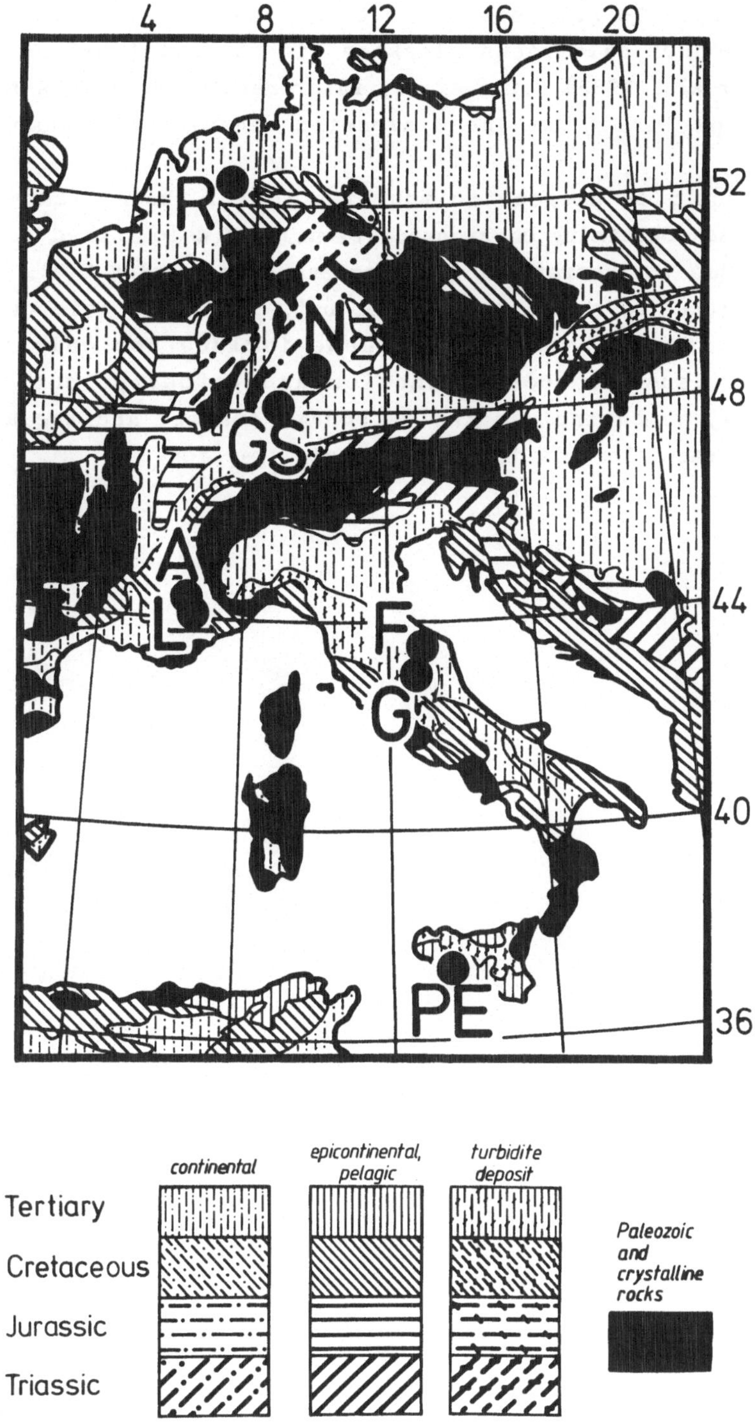

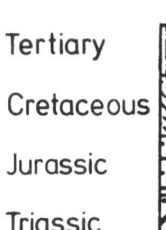

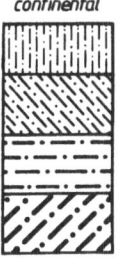

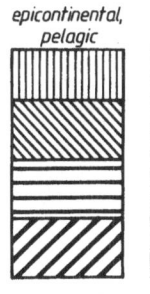

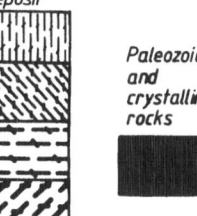

4
8
12
16
20
52
48
44
40
36
R
N
GS
A
L
F
G
PE
continental
epicontinental, pelagic
turbidite deposit
Tertiary
Cretaceous
Jurassic
Triassic
Paleozoic and crystalline rocks

and Tertiary. No sections older than Upper Jurassic were chosen, because modern calcareous plankton did not exist before this age (SCHOLLE et al., 1983). Therefore, interpretations in this study can be based partially on the diagenetic behavior of modern pelagic sediments.

Sections with diagenetic mobilization of the silicate fraction (e.g., those containing chert concretions) and dolomitized zones were not used due to theoretical restrictions (see section 2.2). Analyses were restricted to relatively simple, but, for diagenetic questions, very promising parameters. These include not only physical parameters, such as rock compaction and porosity, but chemical parameters, which include the carbonate content and the amount of minor and trace elements in the carbonate fraction.

2 METHODS FOR THE QUANTIFICATION OF DIAGENETIC CARBONATE DISSOLUTION AND CEMENTATION PROCESSES

Up to this point, methods for quantifying diagenetic processes in order to provided a numerical evaluation of cemented and dissolved carbonate contents have scarcely been developed. The theoretical experiments for calculating the amounts of pressure dissolution and cementation in various types of sphere packing models should be cited (RITTENHOUSE, 1977a,b; MANUS & COOGAN, 1974; VINOPAL & COOGAN, 1978; MITRA & BEARD, 1980) as well as those attempts to evaluate the cement content of skeletal carbonates in acetate peels and thin sections (COOGAN, 1970; MEYERS & HILL, 1987).

Since the marl-limestone alternations in this study are completely micritic, and calcareous particles are not spheroidal, other methods for quantifying diagenesis (Fig. 3) are suggested which depend on the measurement of three basic parameters. These are carbonate content, rock porosity, and compaction (section 2.1). They are mathematically related in the derived carbonate compaction law (section 2.2). The carbonate compaction law describes sediment or rock composition in relation to both degree of compaction and porosity. According to this law, the noncarbonate content of the original volume is calculated which provides evidence of the primary compositional variations. Thereafter, a carbonate mass balance calculation is performed based on mathematical decompaction (section 2.3). The mass balance calculation gives the mean primary sediment composition of the existing marl and limestone layers and the amounts of dissolved and cemented carbonate. Compaction and known porosity-overburden data determine the porosity at the onset of cementation and the timing of lithification (section 2.4).

The above-mentioned measurements and calculations were carried out in section 3 for ten different profiles of marl-limestone alternations. The resulting data were combined in sections 4 and 5 to develop general models. Simple and somewhat imprecise methods (section 6) are applicable to diagenetic mass exchanges. Mass balance calculations and other methods to determine the amount and

QUANTIFICATION OF DIAGENETIC PROCESSES

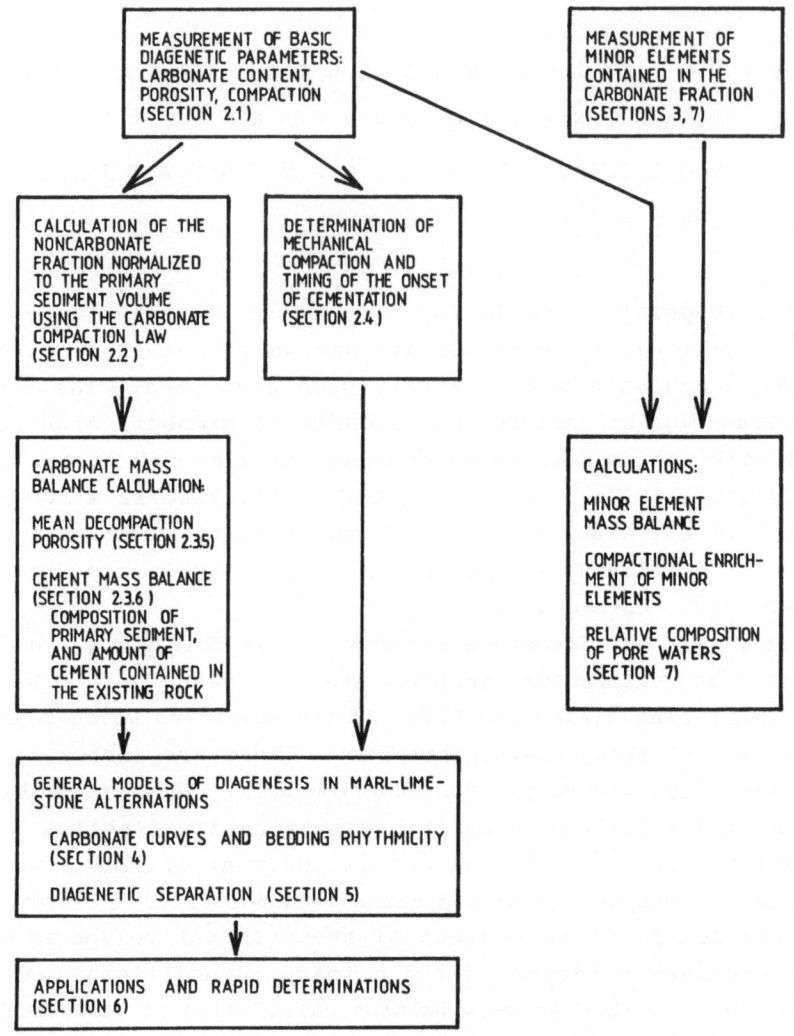

Fig. 3 Methods used in order to quantify diagenetic carbonate dissolution and cementation processes.

redistribution of minor elements in the carbonate fraction are presented in section 7. The primary depositional processes of marl-limestone alternations are discussed in only one example - from the extremely well-bedded Upper Jurassic in southern Germany (section 8).

For the most important calculations instructions and programs are available (RICKEN, 1985b) using Texas Instruments calculators (TI 58, 59). These programs calculate: density and porosity of carbonates from laboratory data; carbonate and minor element mass balances; porosity, compaction, and time in sediment piles with increasing overburden; carbonate curves in limestone layers; and development of both the carbonate content and compaction with decreasing porosity in marl layers due to cement integration and cubic sphere packing models (Figs. 66a, 67a).

2.1 Evaluating Basic Diagenetic Parameters

2.1.1 Carbonate Content and Porosity

Both carbonate content and porosity can be easily evaluated. After a reaction time of 20 minutes using warm 1 M HCl, the carbonate content, expressed as $CaCO_3$, was quantitatively determined by titration of excess HCl using a solution of 0.5 M NaOH. The relative error is $\pm 0.02\%$.

Rock porosity was evaluated by comparing the weight of a given sample in air versus that in water (Archimedes principle) and using its mineral density. The mineral density was obtained from the specific densities of the carbonate and noncarbonate fractions (2.71 and $2.75 g/cm^3$, respectively) and from the proportions of those fractions in the sample studied. The samples were dried for 24 hours at 80°C then cooled in a desiccator and coated with a thin layer of shellac. Errors resulting from the shellac coating were corrected.

2.1.2 Compaction

Many studies of clay and shale compaction have been conducted. In addition to the previously mentioned theoretical considerations (in RITTENHOUSE, 1977a,b; MANUS & COOGAN, 1974; VINOPAL & COOGAN, 1978; MITRA & BEARD, 1980), other experiments and measurements of porosity reduction due to overburden have been carried out. The best presentation of the problem and compilation of porosity-overburden curves in shales were given by BALDWIN (1977), RIEKE & CHILINGARIAN (1974), PERRIER & QUIBLIER (1974), ROLL (1974), and BALDWIN & BUTLER

(1985). Although some limestones display little compactional deformation (PRAY, 1960; STEINEN, 1978; BATHURST, 1980a,b), data from the Deep Sea Drilling Project (DSDP), porosity measurements from drill holes, and compaction experiments (SHINN et al., 1977; BHATTACHARYYA & FRIEDMAN, 1983; 1984) provide clear evidence that carbonate compaction does indeed occur, especially in calcilutites (SCHLANGER & DOUGLAS, 1974; HAMILTON, 1976; SCHOLLE, 1977; LOCKRIDGE & SCHOLLE, 1978; SCHMOKER & HALLEY, 1982; KOPF, 1983).

Systematic evaluation of compaction in carbonates is yet to be done. There are several observations of deformed fossils, ooids, and similar structures. Compaction varies between a few percent to 90% of the original thickness, depending mainly on the overburden and the carbonate content (e.g., EINSELE & MOSEBACH, 1955; KAHLE, 1966; WOLFE, 1968; ZANKL, 1969; JANOWSKY, 1970; BALDWIN, 1971; KETTENBRINK & MANGER, 1971; CHAND et al., 1977; BEACH & SCHUMACHER, 1982; BATHURST, 1983; MEYERS & HILL, 1983).

In this study, the evaluation of compaction is based mainly on the amount of deformation found in originally cylindrical burrow tubes. Burrow deformation (D), expressed as percentage loss of original diameter, equals:

$$D[\%] = 100 - \frac{100b}{a} \quad , \tag{1}$$

where (a) is the length of the major axis and (b) the length of the minor axis of the deformed burrow ellipse. The axes are perpendicular to the length of the burrow tube; the major axis must lie parallel to bedding if the paleoslope was not very steep. It is assumed in eq. 1 that the length of the major axis remained constant during the burrow deformation (Fig. 4). Burrow deformation occurs either in the soft or semilithified sediment due to the mechanical compaction of the pore space or at the various sites of pressure dissolution (e.g., microseams) within the grain structure. If the major axis was laterally stretched during compaction, sediments in aseismic basins should become more and more folded when they are subject to increasing overburden!

In order to avoid oblique cross sections of the burrow tubes, well-preserved burrows were extricated from the rock. The axes were then measured using a mm scale. Burrows perpendicular to bedding were used to test whether or not the tubes were originally cylindrical.

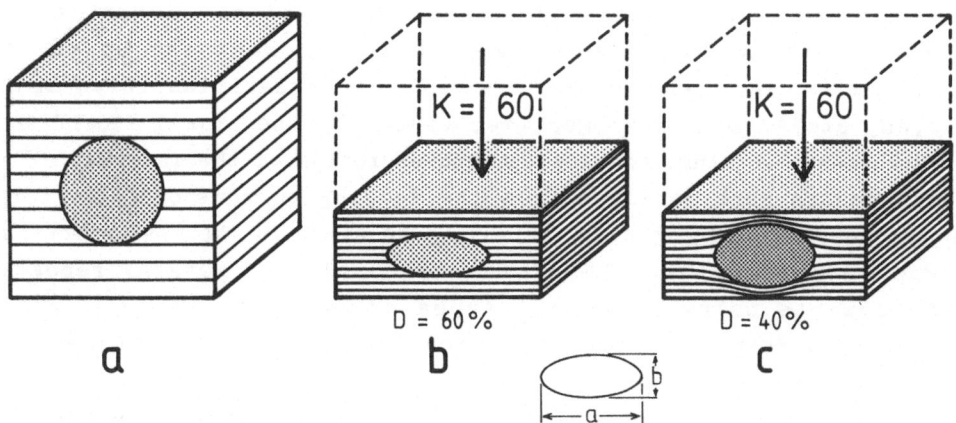

Fig. 4 How to evaluate rock compaction (all cross-sections cut perpendicular to both bedding and burrow tube).
a: Primary volume of sediment containing a circular, sediment-filled burrow.
b: It is assumed that in most cases the degree of compaction (K) is proportional to the degree of burrow deformation (D).
c: Burrows which are cemented early in diagenesis show less deformation (D) than the actual compaction of the total rock (K). An example of indirect evaluation of rock compaction is given in the text.

Direct Evaluation of Compaction

As PLESSMANN (1977) and CRIMES (1975) pointed out, bioturbation structures are highly suitable for the evaluation of compaction, since they are a part of the sediment itself. Contrary to the compactional behavior of fossils, which can completely or partly resist compaction, bioturbation structures usually deform to the same degree as the surrounding sediment. Thus, the degree of rock compaction (expressed as percentage loss of the original sediment volume) is commonly proportional to the degree of burrow deformation (Fig. 4A,B).

Except for the upper 3 to 10cm of sediment, which is often completely mixed due to intense bioturbation, trace fossils used in this study came from the upper meter of sediment (BERGER et al., 1979; EKDALE et al., 1984). Modern trace fossils commonly found in carbonates (KENNEDY, 1975) occur in five levels of bioturbation (WETZEL, 1981). The trace fossils with originally circular burrow tubes, which were used in this study, all lie in levels III (Planolites) and IV (Chondrites, Teichichnus, and Thalassinoides; HÄNTZSCHEL, 1975); level III overlies level IV.

Repeated measurements show that evaluation of compaction is inaccurate. Errors resulting from the inexactness of burrow measurements, the degree of preservation, irregularities (branching, curves, etc.), and inhomogenities caused by the animal itself from displacement, sorting, compaction, and excretion of the material. The best results were obtained from the feeding burrows of <u>Planolites</u>, <u>Chondrites</u>, and <u>Teichichnus</u>. Usually, the error in the measurement of compaction is below $\pm 10\%$ of the mean value of several repeated measurements using different types of burrows. Therefore, if possible, a mean value of compaction was calculated from several single measurements.

If adequate bioturbation structures are not available, compaction measurements can be carried out using "steinkerns" of formerly aragonitic shells, since the hard shells often get dissolved during early burial (SEILACHER et al., 1976). For this purpose, sediment-filled living chambers of straight or coiled ammonites are suitable; one solely has to know the original ratio of axes, for instance, from the slightly-deformable calcitic aptychi. These measurements are usually equivalent to those obtained from deformed bioturbation structures.

Very resistant calcite shells and massive grains, such as ooids, often deform while maintaining a constant volume. Their deformation cannot be described by eq. 1. RAMSEY & HUBER (1983) give several examples to calculate these types of deformations.

Indirect Evaluation of Compaction

Very often it is impossible to directly ascertain the degree of compaction. The ability to perform this calculation depends on the amount of early cementation of certain burrows and fossils as compared to that of the surrounding rock. In early, selectively cemented burrows, the degree of burrow compaction is less than the compaction of the total rock, thereby representing only the amount of compaction prior to the onset of cementation inside the burrow (K>D, Fig. 4C).

However, the total rock compaction can be determined from the degree of partial compaction of selectively cemented burrows or shelly fossils using the compaction law given in section 2.2. Calculations are based on the simple assumption that the burrow infill and the surrounding sediment consist of the same material; thus the absolute clay content in the burrow and in the neighboring sediment should be the same (the absolute clay content is normalized to the original sediment volume, see the following section). Once one obtains the absolute clay content of the burrow and the carbonate content and porosity of the surrounding rock, the total rock compaction can be calculated by solving eq. 4.

Example (Fig. 4C, rock porosity is here neglected): A selectively cemented burrow tube lost 40% of its original thickness and has a carbonate content of 83.3% while the surrounding rock has only 75% $CaCO_3$. The resulting absolute clay content of the burrow amounts to 10% (compaction law, eq. 3). Since the carbonate content of the neighboring rock is 75% (and if the absolute clay content of the sediment was the same as for the burrow), then the loss of volume in the surrounding rock is 60% (eq. 4).

In spite of simplifications, indirect evaluation of the total rock compaction does give useful results (Table 1). Errors arise from compositional differences between burrow or steinkern infill and the neighboring sediment. However, such errors can be partially compensated for when deformation measurements are carried out more accurately with well-preserved, selectively cemented burrows rather than with non-cemented bioturbation tubes.

Table 1: Difference between direct and indirect determination of compaction. In the calculations, the existing amount of rock porosity was ignored.

Sample		Carbonate content		Compaction			Difference between the measured and calculated amount of compaction in the rock matrix
		rock matrix	selectively cemented burrows	sel. cem. burrows measurement	rock matrix, measurement	rock matrix, calculated	
Gubbio	166	47.43	87.94	43.0	80.0	86.9	-6.9
1	181	77.90	83.44	35.0	50.8	51.3	-1.3
Gubbio	62	86.20	90.24	58.1	84.5	70.4	14.1
2	72	78.68	92.69	47.8	80.0	82.1	-2.1
	75	65.37	87.30	54.4	82.6	83.3	-0.7
Logis	6	74.76	86.11	36.0	71.0	64.8	6.2
du Pin	40	69.63	83.93	47.0	75.7	71.9	3.8

2.2 Derivation of the Compaction Law

The "carbonate compaction law" describes the general relationship between carbonate content, compaction, and porosity (the basic diagenetic parameters). The relationship is valid for any given calcareous sediment or rock sample, independent of the specific lithology and irrespective of whether or not the diagenetic carbonate system is closed. The carbonate compaction law stipulates that during diagenesis the original values of the carbonate fraction and the pore solution should be diagenetically mobile and that the original volume of noncarbonate solids should be essentially immobile. Thus, the original volume or absolute amount of the noncarbonate fraction should be a constant factor during carbonate diagenesis (the noncarbonate fraction consists mainly of clay minerals and silt-sized silicates). This does not exclude, however, that individual constituents of the noncarbonate fraction (e.g., clay minerals) may undergo isochemical alterations during diagenesis.

In fact, these conditions are found in many rocks composed of carbonate and clay (WEDEPOHL, 1970). Except for the mobilization of opal, which can be recognized in the form of chert concretions (and thus be avoided), the formation of pyrite (BERNER, 1984), and the dewatering of clay minerals during high overburden (BOLES & FRANKS, 1979), diagenetic changes of the noncarbonate fraction can be usually neglected. Often the noncarbonate fraction in the marl beds and in the limestone layers have similar if not the same mineralogical composition (SUJKOWSKI, 1958; HÖLLER & WALITZI, 1965; BAUSCH et al., 1982; BURGER, 1982; POLLASTRO & MARTINEZ, 1985). Supposedly, the carbonate compaction law can be applied to more than 90% of existing sediments or rocks consisting of carbonate and clay.

Derivation (Fig. 5): The carbonate compaction law explains the diagenetic changes for a given calcareous sediment or rock sample in terms of volume. Original calcareous sediment consists of carbonate grains (C_o), noncarbonate grains (NC), and water-filled pores (n_o). Most of these parameters become absolutely and relatively altered during diagenesis: The original volume decreases due to mechanical and chemical compaction (K) and the original pore space decreases to provide the present rock porosity (n). Moreover, the original carbonate volume can either increase or decrease due to cementation or dissolution; this gives the carbonate content of the existing rock (C). The noncarbonate volume (NC) changes relative to the changing amount of carbonate. However, during diagenesis, <u>the noncarbonate content remains constant if it is expressed as a percentage of the primary sediment volume</u> (or "NC_d," the noncarbonate fraction normalized to the decompacted, primary sediment volume, Fig. 5). Hereafter, the standardized noncarbonate content will often be referred to simply as the "absolute clay content". Commonly, it is a constant factor during carbonate diagenesis.

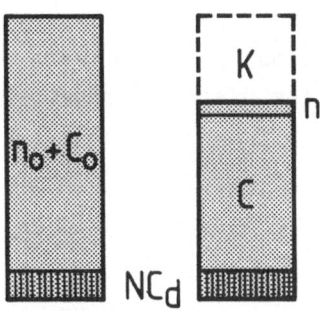

Fig. 5 Relationship between primary sediment (left) and resulting carbonate rock (right) expressed in terms of volume. K=compaction, n=porosity, C=carbonate fraction, NC_d=standardized noncarbonate fraction. During diagenesis, the NC_d fraction remains constant only when it is expressed as a percentage of the primary sediment volume.

If the primary sediment volume is 100%, the compacted volume (of the sediment or rock) is (100-K). The amount of solids in the compacted volume is (100-n); this equals (100-K)((100-n)/100), when related to the original sediment volume. Again, the solids consist of both the carbonate volume (C) and the noncarbonate volume (NC), where NC is (100-C) of the pore-free solids. The following expression is derived when NC is expressed as a portion of the original sediment volume (NC_d):

$$NC_d[vol\%] = \frac{(100-K)(100-n)(100-C)}{10000} \quad , \qquad (2)$$

where K is the percentage of compaction of the primary sediment volume, n is the porosity expressed as a percentage of the sediment or rock volume, and C is the carbonate volume expressed as a percentage of the volume of pore-free solids.

Eq. 2 is the general form of the carbonate compaction law. It is named the carbonate compaction law because in dense, lithified rocks with zero porosity the law relates only carbonate content to compaction. Hence, the standardized noncarbonate fraction of dense or nearly dense rocks (NC_{dr}) is

$$NC_{dr}[vol\%] = \frac{(100-K)(100-C)}{100} \quad , \quad (3)$$

(this is the special form of the compaction law). If one solves equation (2) for compaction (K), carbonate volume (C), and porosity (n) one gets the following basic equations:

$$K[vol\%] = 100 - \frac{NC_d}{(1-0.01n)(1-0.01C)} \quad (4)$$

$$C[vol\%] = 100 - \frac{NC_d}{(1-0.01K)(1-0.01n)} \quad (5)$$

$$n[vol\%] = 100 - \frac{NC_d}{(1-0.01K)(1-0.01C)} \quad (6)$$

As already shown, the compaction law deals with volumes which are independent of the size and distribution of grains and pores. The compaction law alone does not enable one to distinguish different types of carbonate (e.g., cement and primary carbonate). Moreover, no distinction is made between the different types of compaction (such as mechanical versus chemical compaction, section 5.4). Since the specific weights of calcite and the nonsoluble residue are often very similar (about $2.7 g/cm^3$), the expression of the percentage of weight (carbonate content data resulting from analyses) as a percentage of volume (as used in the compaction law) is usually not necessary. For the most part, errors resulting from these slight fluctuations can be neglected.

The porosity of lithified carbonates can be largely ignored; therefore, calcareous rocks can usually be evaluated with the special form of the compaction law (eq. 3). When porosity is insignificant and the absolute clay content is constant, the carbonate compaction law shows a nonlinear relationship between carbonate content and

CARBONATE COMPACTION LAW

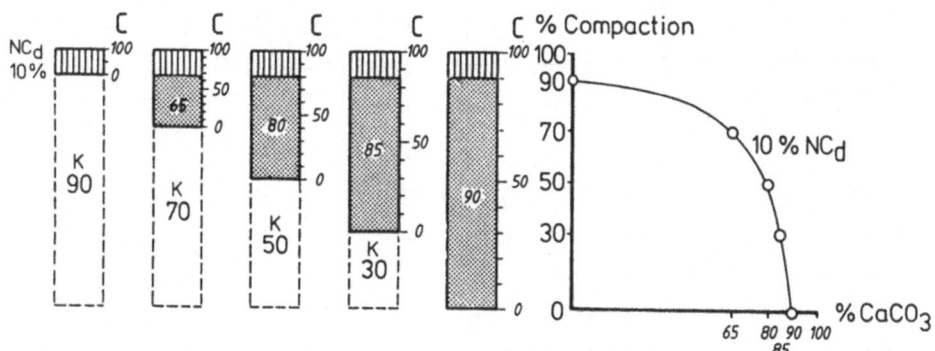

Fig. 6 Simplified derivation of the compaction law, neglecting porosity. Example resulting from chemical compaction: A carbonate rock with 90% CaCO₃ is chemically compacted to K=30, 50, 70, and 90% of the original thickness. Carbonate content (C) diminishes nonlinearly, because the absolute clay content (NC_d=10%) remains constant during diagenesis.
<u>Note</u>: The same data result if the primary sediment (containing the same absolute clay content) first undergoes mechanical compaction and subsequently complete cementation.

compaction (Fig. 6). This relationship changes somewhat if the present porosity is below 30% (Fig. 7). The influence of pore volume is clearly recognizable only when the porosity is high (see Fig. 65).

1. Curves from the compaction law representing carbonate content and the degree of compaction for various absolute clay contents (Fig. 7) are approximately parallel to the $CaCO_3$ axis when compaction is high; whereas when carbonate content is high, the curves are approximately parallel to the compaction axis. It becomes clear that calcareous rock can still have a high carbonate content, in spite of strong compaction. For example, typical limestones in this study have absolute clay contents between 2.5 and 10%. A limestone with an absolute clay content of $NC_d=5\%$ would correspond to a primary carbonate content of $C_o=83.3\%$ if we assume a typical primary porosity of 70%. If this sediment becomes completely cemented without suffering compaction, the carbonate content would be 95% (Fig. 7). If the sediment becomes mechanically compacted by 50% and then cemented, the carbonate content would be 90%.

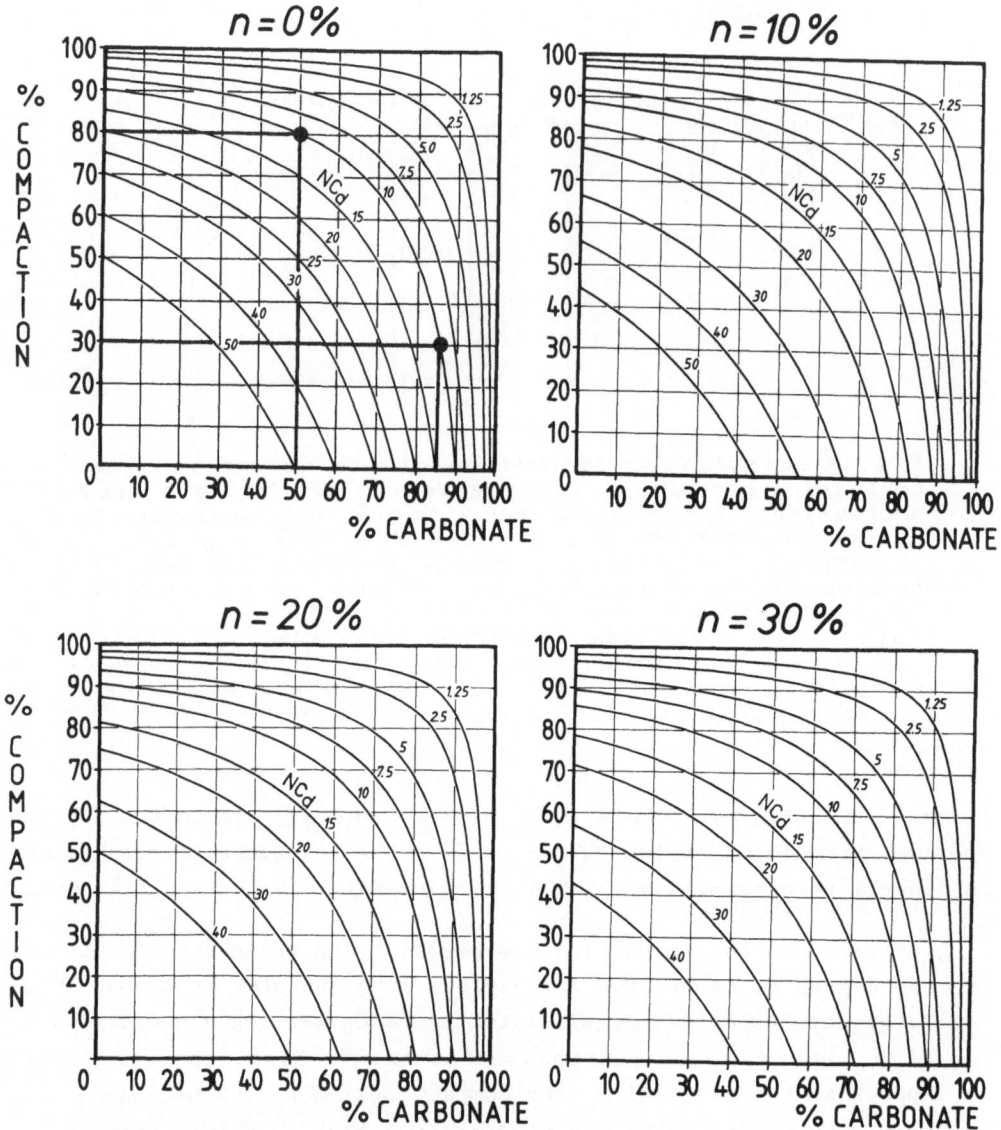

Fig. 7 Relationship between compaction and carbonate content for different absolute clay contents (NC_d) and porosities (n) calculated after the compaction law.
<u>Examples</u>: They show dense calcareous rocks (n=0%) with absolute clay contents of 10%. Sample containing a carbonate content of approximately 85% displays a compaction of 30%. However, sample containing a carbonate content of 50% shows a compaction of 80%.

After 80% mechanical and chemical compaction the resulting rock would still have 75% $CaCO_3$ if its porosity is low. The carbonate content would decrease considerably, but not before compaction became extreme (e.g., 90% compaction, 50% $CaCO_3$, 0% porosity).

2. Carbonate content and compaction measured in marl-limestone alternations should plot on one of the absolute clay curves presented in Fig. 7, if the existing alternation is mainly generated from a uniformly-composed, basic substance containing a constant or nearly constant absolute clay content. Nevertheless, one cannot expect that the paths of the theoretical curves will be completely identical to those produced by field data. This is due to primary compositional variations and due to the compactional history of the sediment (e.g., the amount of mechanical compaction).

2.3 Carbonate Mass Balance and Primary Sediment Composition

Since the compaction law alone does not calculate the movements and amounts of dissolved and cemented carbonate, a mass balance calculation is carried out which is based on the condition that the diagenetic carbonate system remained closed within an interval of several meters. The carbonate mass balance calculation (Fig. 8) relies on mathematical decompaction in order to reconstruct the original composition and the amount of carbonate redistribution in both marl beds and limestone layers. During decompaction, reconstructed limestone sediment has too low a primary porosity due to cementation, while the reconstructed marl sediment has a relatively high porosity due to carbonate dissolution. The mass balance calculation redistributes the carbonate content between dissolution-affected and cemented zones (i.e., marl beds and limestone layers, respectively), so that the same primary pore space is obtained for reconstructed sediments of both zones (Fig. 8). If the carbonate system was closed during diagenesis, the mass balance calculation determines the amounts of the different carbonate fractions (primary, cemented, dissolved, and relic carbonate).

SCHEME OF CARBONATE MASS BALANCE CALCULATION

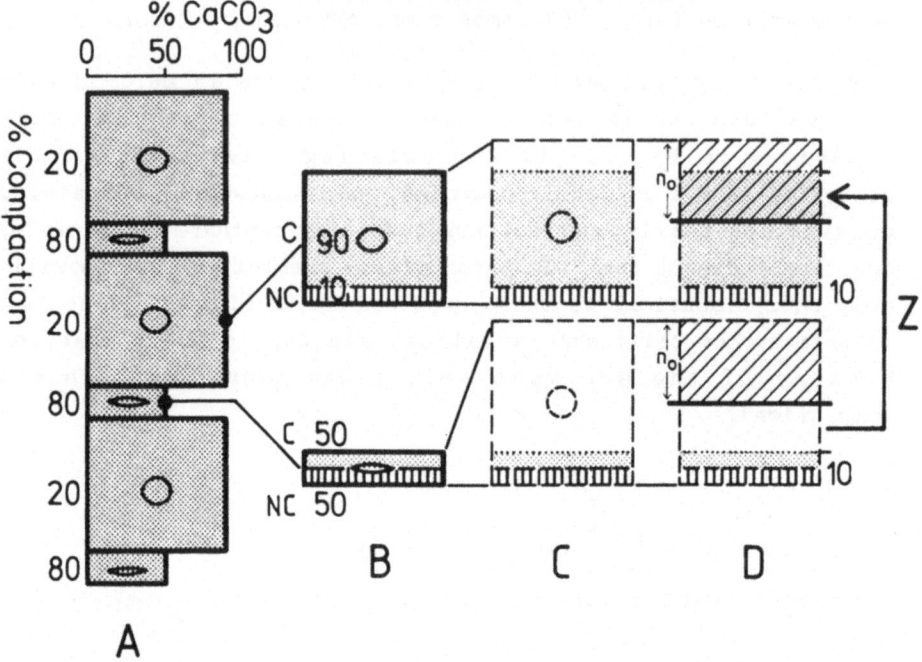

Fig. 8 Simplified, schematic diagram representing the carbonate mass balance calculation.
A: Marl-limestone alternation. Limestone layers: 90% $CaCO_3$, 20% compaction. Marl layers: 50% $CaCO_3$, 80% compaction. Degree of compaction can be recognized by the deformation of cylindrical burrows.
B: Separate description of two beds, showing clay content (NC, vertically striped areas) and carbonate content (C, shaded areas).
C: Calculated decompaction of a limestone layer and marl bed. The deformed burrow again appears circular. The decompacted limestone layer shows too low an original pore space, while the decompacted marl layer has a relatively high primary porosity.
D: If the primary pore space in both layers is equal (n_0, hatched areas), the reconstructed marl layer sediment has an additional amount of porosity due to carbonate dissolution (n_0 is the average decompaction porosity of both layers). In the reconstructed limestone layer sediment, the dissolved carbonate from the marl layer fills a part of the primary pore space due to cementation (Z).

2.3.1 Closed or Open Carbonate System
During Burial Diagenesis

The carbonate mass balance calculation gives realistic results only if carbonate is not brought into or removed from the studied sequence during diagenesis (that is, a closed system for carbonate). The still prevalent opinion that carbonate diagenesis usually occurs in an open system with the participation of huge volumes of migrating pore solutions (PRAY, 1966; DUNHAM, 1969; BATHURST, 1976; MORROW & MAYERS, 1977) has been refuted by geochemical arguments (VEIZER, 1978; BAKER et al., 1982). In contrast to this, PINGITORE (1976, 1982) and BRAND & VEIZER (1980) have proposed the concept of "partly closed reaction zones". According to this concept, the diagenesis occurs in micropores predominantly independent of the pore solution in the macropores, thus requiring less pore fluid. Nevertheless, BRAND & VEIZER still presume that minor element diagenesis occurs primarily as a result of reaction with meteoric water. BAKER et al. (1980) and ELDERFIELD & GIESKES (1982) demonstrated, however, that removal and enrichment of trace elements can also occur under marine pore water conditions. Moreover, numerous data are available concerning lithification of pelagic carbonates (e.g., SCHLANGER & DOUGLAS, 1974; SCHOLLE, 1977; GARRISON, 1981). Authors usually agree that in deep-sea carbonates and subaerial chalks, cement is generated within the sediment column due to nearby dissolution-precipitation processes. MATTER (1974) characterized these processes as "autolithification".

Isotopic data ($\delta^{18}O$, $\delta^{13}C$) from many limestones point to lithification under marine conditions in a closed carbonate system (HUDSON, 1977; CZERNIAKOWSKI et al., 1984). For this redistribution, the long distance transport of enormous masses of carbonate cement are no longer required. Moreover, authors are currently more willing to accept mechanical compaction of carbonates. This reduces the pore space to be cemented (e.g., mean amounts of cement calculated in this study are approximately one third of the total carbonate in the limestone layers). Comprehensive discussions of this problem are given by SHINN et al. (1977) and BATHURST (1980a,b).

The following results confirm the opinion that diagenesis in marl-limestone alternations occurred in a closed or nearly closed carbonate system:

1. As already observed from HARMS & CHOQUETTE (1965), BUXTON & SIBLEY (1981), MERINO et al. (1983), and KOEPNIK (1985), carbonate is reprecipitated close to stylolitic seams. In the Upper Oxfordian marl-limestone alternation in southern Germany, cementation occurred preferentially above and below the stylolitic bedding planes (Fig. 9), suggesting transport by diffusion with and against the compaction flow (EINSELE, 1977; WEDEPHOL, 1979; PINGITORE, 1982; see section 5.1).

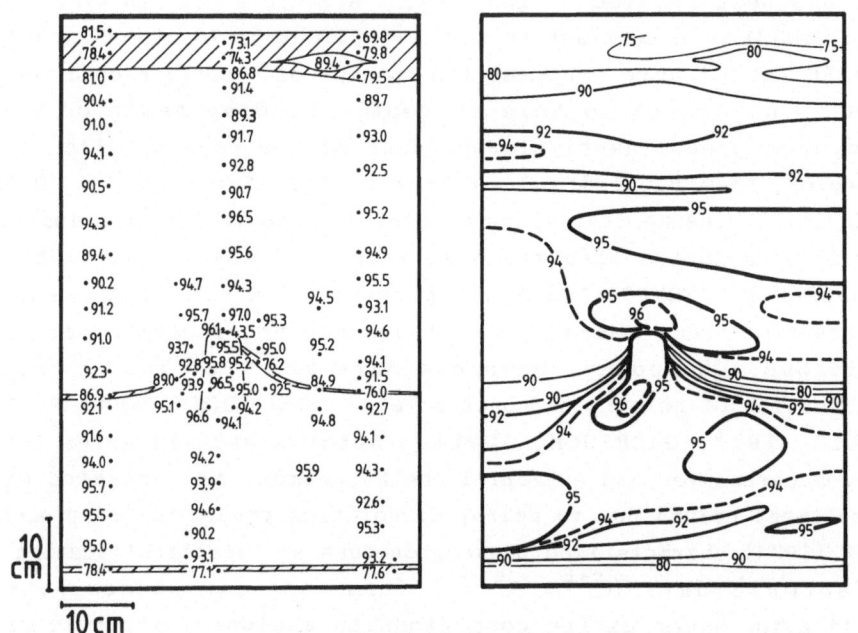

Fig. 9 Carbonate content in the neighborhood of a diagenetic stylolite. The highest carbonate content (96%) was measured directly above and below the stylolite. Neuffen Quarry, Swabian Alb, Upper Oxfordian (see Fig. 30D).

2. Mathematical decompaction of the 10 sections studied (each several meters thick) yields mean primary porosities between 65 and 80% for pelagic to hemipelagic alternations and about 60% for the more neritic profiles of the Rheine and Logis du Pin sections (Fig. 10). Porosities of a similar order of magnitude were observed in modern fine-grained carbonates from shelf and deep-sea environments (KELLER & BENNETT, 1970; HAMILTON, 1976; KELLER et al., 1976; MAYER, 1980). If substantial carbonate transport into

the systems of the studied sections had occurred, one would obtain smaller decompaction porosities. On the other hand, if larger amounts of carbonate had been released, resulting porosities would be larger.

MEAN DECOMPACTION POROSITY[%]

Fig. 10 The mean primary porosity in the sections studied, which is calculated from decompaction of the entire section. Stars indicate the decompaction porosities from the relatively nearshore sections Logis du Pin (southern France) and Rheine (northern Germany).

2.3.2 Uncertainties Resulting from the Method

The method of balancing the cement content contains several sources of error which reduce the possibility of yielding precise results of the primary sediment composition. One source of error results from calculating the mass balance with a mean value of primary (decompacted) porosity, although small differences in porosity must have actually existed in the original sediment.

While KELLER & BENNETT (1970) gave only a 2% difference in porosity between pelagic carbonates, terrigenous material, and red deep-sea clay (mean porosities: 72%, 73%, 74%, respectively), HAMILTON (1974, 1976) gave porosity differences of 9% (72% for calcareous and terrigenous ooze, 81% for red, deep-sea clay). OSMOND (1981) cited differences of 14% (67% for calcareous ooze and siliciclastic material, 81% in silt and clay).

However, one can expect that the maximum porosity differences of 9 to 14% were not present in the primary sediment, since lithified marl-limestone alternations (after diagenesis) usually have carbonate differences between 30 to 50%. Therefore, one can assume that the original fluctuations in porosity were below 5% (see the prediagenetic alternation of Porto Empedocle, Italy, section 3.1). Taking into account porosity differences of this order of magnitude (n_o = 58 to 78% $\pm$2.5%, NC_d = 3 to 11%) the maximum uncertainties in evaluating primary carbonate differences must be between $\pm$0.5 and $\pm$5.8%. The degree of uncertainty in the primary composition will be given for every balance calculation carried out (section 3.2 to 3.5).

Another uncertainty results from inexactness of the previously mentioned measurements of burrow deformation. According to this author's experience, usually more than 30 evaluations of compaction per section are necessary to obtain representative average values in the balance calculations.

2.3.3 Sampling Procedure

The thickness of the sections should be chosen such that at least 30 evaluations of compaction within successive marl and limestone layers can be made. The probability of getting a closed carbonate system is higher in thicker sections. Sections used in this study are only 3 to 15 meters thick and comprise, on the average, 28 carbonate oscillations.

In order to apply the carbonate compaction law, one has to determine the carbonate content and porosity of the rock matrix at every site where compaction was measured. If evaluation of compaction is carried out indirectly (see section 2.1.2), the carbonate content in selectively-cemented burrows or steinkerns also has to be determined. Moreover, a tight sampling should guarantee that every carbonate variation in the section is completely recorded.

In the studied sections, rock porosity is below 15%. According to the compaction law, the influence of rock porosity on the calculations is small. After several porosity determinations, rock porosity can be expressed in relation to carbonate content using regression curves. Then porosity can be indirectly evaluated.

2.3.4 Testing Primary Compositional Differences

The absolute clay content, which has to be standardized to the primary volume of sediment, is calculated from measured data using eq. 2. If no or only slight diagenetic mobilization of the noncarbonate fraction occurred (see section 2.2), then the absolute clay content is a measure of the variations in the amount of the noncarbonate fraction in the primary sediment. However, individual values of the absolute clay content only provide weak evidence for primary variations in composition, since compaction cannot be measured accurately. If the mean absolute clay content from various measurements remains unchanged in the present marl and limestone layers and the original porosity was constant throughout the primary sediment, then the primary carbonate content was also constant and present $CaCO_3$ oscillations must form predominantly due to diagenetic carbonate redistribution. However, if the primary sediment had significantly large variations in porosity, then the reconstructed primary sediment would have carbonate variations even if the absolute clay content was constant in the studied sections. According to section 2.3.2 and to the prediagenetic sequence of Porto Empedocle (section 3.1), primary porosity variations in the studied alternations were in all probability, below 5%. Thus, the determination of the standardized noncarbonate content provides the first powerful tool to evaluate primary compositional differences (both clay and carbonate).

2.3.5 Evaluation of the Mean Primary Porosity
Resulting from Decompaction

The mass balance calculations are carried out in two major steps: First, the mean original porosity (based on mathematical decompaction) is calculated (Fig. 11a,b,c), and second, the content of dissolved and cemented carbonate is determined using mass balance calculation (Fig. 11d).

Fig. 11a: The carbonate oscillations found in the studied section are transformed into a histogram representing the amount of rock in the total section which contains a given amount of $CaCO_3$. For this purpose, the carbonate curve of the section, which is very carefully determined (section 2.3.3) has to be graphically integrated for individual $CaCO_3$ intervals. Usually, 10 to 20 carbonate intervals are formed.

Fig. 11b: Each column of rock with a given carbonate content must be divided into mean volume proportions of rock porosity (n), carbonate (C), and noncarbonate content (NC).

CARBONATE MASS BALANCE

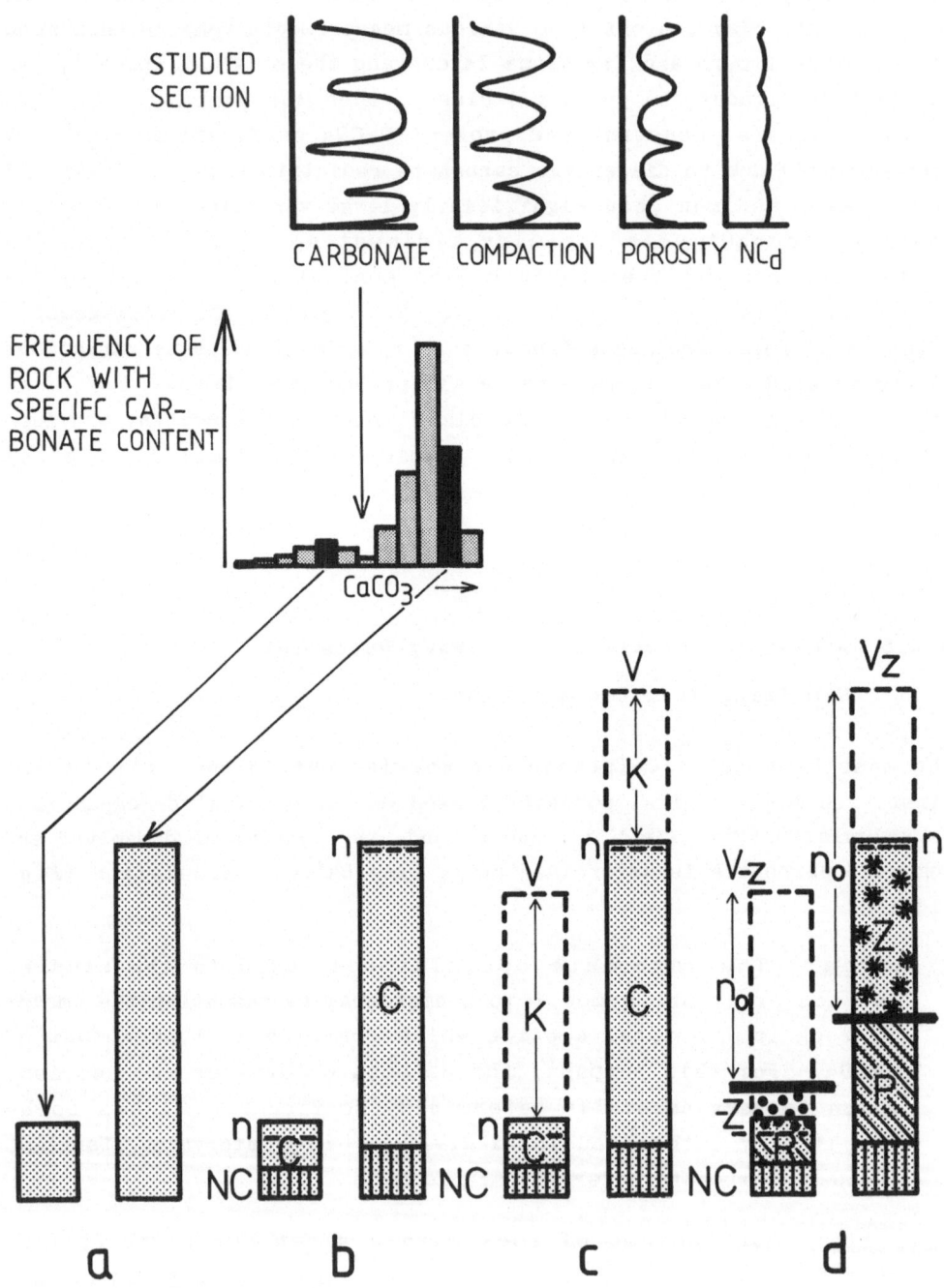

STUDIED SECTION

CARBONATE   COMPACTION   POROSITY NC$_d$

FREQUENCY OF ROCK WITH SPECIFC CAR- BONATE CONTENT

CaCO$_3$

a   b   c   d

<u>Fig. 11c</u>: The amount of compaction for every carbonate class in
the histogram is determined using mean values of carbonate
content, absolute clay content and porosity (eq. 4).
Determination of the amount of compaction allows one to decompact
the height of the existing rock column (h) by calculating the
original rock column thickness (h*) with

$$h^* = \frac{100}{100-K} \times h \quad . \qquad (7)$$

The original rock thickness in all carbonate classes has to be summed
to give the original sediment volume (V). If diagenesis occurred in a
closed carbonate system, the difference between the original sediment
volume and the separately added amount of pore-free solids of the rock
result in the mean decompaction porosity (n_o). The mean decompaction
porosity should be equal to the original porosity of the sediment,
because burrows used for the compaction measurement are generated
slightly below the sediment–water interface (see section 2.1.2).

2.3.6 Carbonate Mass Balance Calculation

The amount of dissolved and cemented carbonate and the mean primary
composition of the existing marl and limestone layers are ascertained
from a mass balance calculation.

Fig. 11 Carbonate mass balance calculation. Transformation
of the entire section studied into a histogram representing
the amount of rock volume with a certain carbonate content.
a: Example shows two rock volumes with high and low
carbonate content (shaded).
b: The volumes of noncarbonate (NC), carbonate (C), and
porosity (n) in the given rock volumes.
c: Decompaction calculates the original sediment volume (V)
by using the given amount of compaction (K).
d: Subtraction of the mean primary porosity (n_o, which
results from decompacting the entire histogram) from the
original sediment volume. From this, one obtains the
amounts of dissolved (-Z, dotted), cemented (Z, starred),
relic (R, hatched), and "primary" (P, hatched) carbonate
fractions for the two rock columns.

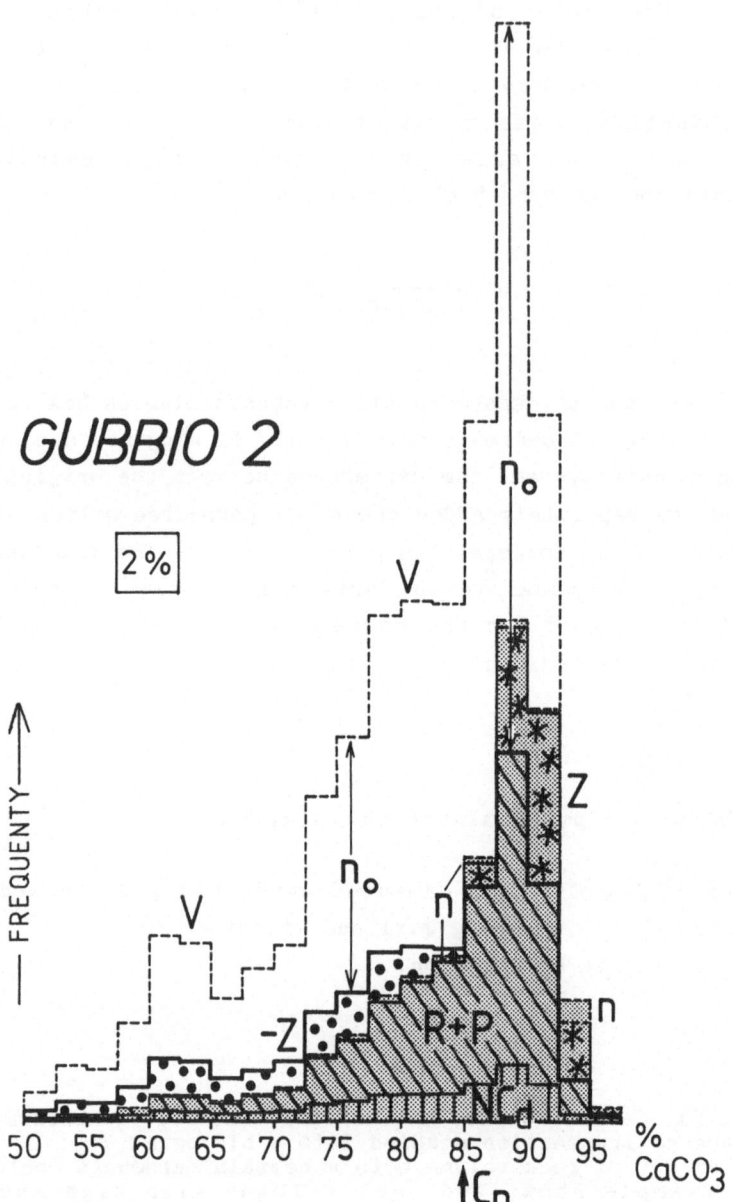

Fig. 12 Carbonate mass balance calculation for the Gubbio 2 section (Italy). The histogram represents a whole-rock balance calculation. V=volume of decompacted sediment, n_o=primary porosity, n=rock porosity, -Z=dissolved $CaCO_3$, Z=cemented $CaCO_3$, R,P=relic or "primary" carbonate, respectively, NC_d=standardized noncarbonate fraction, C_n=$CaCO_3$ neutral value.

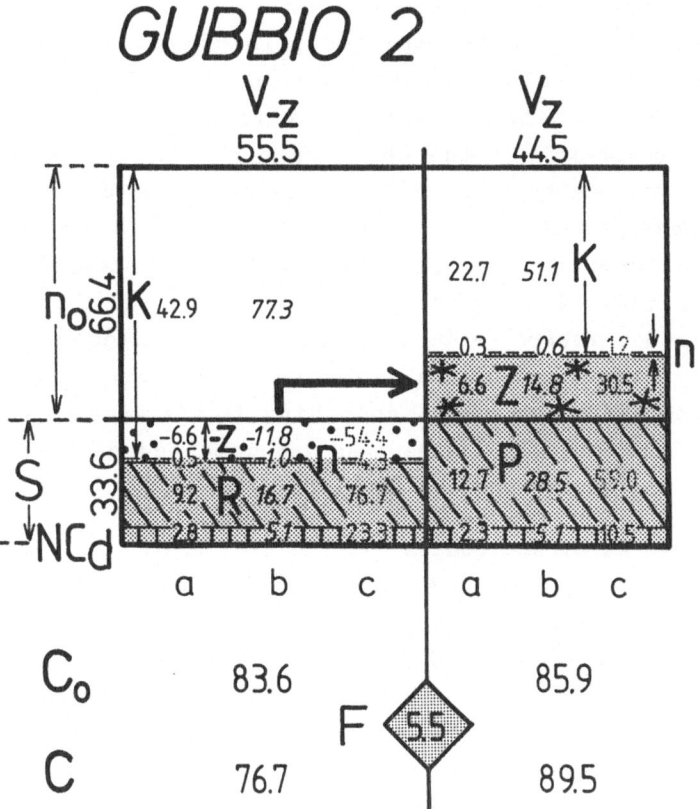

Fig. 13 Box model of a carbonate mass balance calculation
(derived from the histogram in Fig. 12). Outer frame of the
model is equivalent to the decompacted or primary sediment
volume for dissolution (V_{-Z}) and cementation zones (V_Z),
respectively. The existing amount of rock is shaded.
n_o=mean primary porosity, S=original amount of solids,
K=compaction, n=rock porosity, -Z=dissolved, Z=cemented,
R=relic, and P="primary" carbonate, NC_d=standardized
noncarbonate fraction.

Values of K, -Z, Z, n, R, P, and NC_d (small numbers) are
expressed as a percentage of a) the total sediment volume,
b) the volume of dissolution and cementation zones,
respectively, and c) the amount of solids in the
post-diagenetic rocks in both zones.

C_o=mean primary carbonate content of both zones. The
calculation assumes porosity differences of n_o=±2.5% in the
primary sediment (the differences are not shown in the
model). C=existing mean carbonate content for both zones.
F=factor of diagenetic enhancement of primary carbonate
oscillations.

Fig. 11d: The mean decompaction porosity (n_o) has to be subtracted from the decompacted primary sediment thickness for every carbonate class of the histogram. This manipulation reproduces both the amount of solids and pore space in the original sediment. In the cemented parts of the sequence (that is, high calcareous rock in the histogram), the difference between the present and the original amounts of solids correspond to the amount of cement (Z). In the solution-affected portion of the sequence (that is, low calcareous rock in the histogram), the difference corresponds to the amount of dissolved carbonate (-Z). Moreover, the calculation gives the "primary" (P) and relic carbonate fractions (R). According to the mass balance method, the absolute amounts of dissolved and cemented carbonate must be equal in the entire section studied.

For a given carbonate content in the histogram, calculations show neither cement nor dissolved carbonate (an example is the Gubbio 2 section, Fig. 12). This carbonate value is named the "statistically neutral carbonate content" (C_n). If the carbonate system was closed, the neutral carbonate content must be equivalent to the mean carbonate content of the primary sediment. The neutral carbonate value separates the histogram into a zone of carbonate dissolution and one of carbonate cementation. The box model (Fig. 13) is based on the above-mentioned mass balance calculation (Fig. 12). It gives mean values of compaction, porosity and composition for the dissolution and cementation zones in the existing rock and in the calculated primary sediment.

In Fig. 13, the outer frame of the box model represents the primary sediment volume, whereas the shaded part represents the present rock volume. Note that the decompacted volume of layers reduced and enriched in carbonate content (V_{-Z}, V_Z) did not have equal amounts in the primary sediment. Also note that the original sediment underwent different volume changes in both zones due to differential compaction. The mean original carbonate content of dissolution and cementation zones (C_o) is calculated assuming 5% porosity variation in the primary sediment (see section 2.3.2). Thus, the original volume of the pore-free solids (S) in the Gubbio 2 sediment would be 33.6 $\pm$2.5%, indicating a mean primary carbonate content of C_o=83.6 and 85.9%. However, the mean present carbonate content for both zones is 76.7 and 89.5%; thus the original carbonate variations are enhanced by a mean factor of F = 5.5.

2.3.7 Important Definitions Resulting From
The Mass Balance Method

The dissolution and cementation zones as determined by mass balance
calculations do not exactly correspond to the weathering-generated
marl and limestone layers developed in the outcrop. Whether or not
the marl and limestone layers develop depends on the carbonate content
of the weathering boundary (C_w; EINSELE, 1982) which usually is

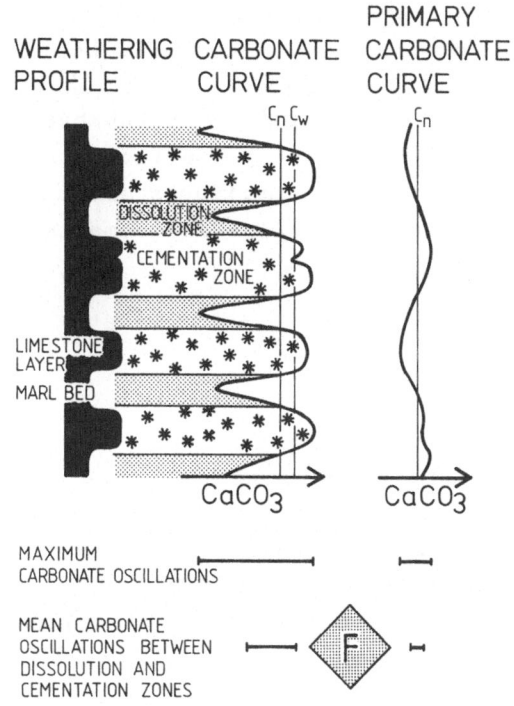

Fig. 14 Definitions for describing marl-limestone
alternations. The transition between dissolution and
cementation zones is defined by the carbonate neutral value
(C_n), while the transition between marl and limestone layers
is defined by the carbonate content at the weathering
boundary (C_w). Also, two ways to describe the carbonate
differences in primary sediment and diagenetically altered
rocks are shown, as well as a method to evaluate the factor
(F) which calculates the diagenetic enhancement between
primary and post-diagenetic $CaCO_3$ differences. Note that in
this study carbonate fluctuations are calculated between the
mean carbonate contents of dissolution and cementation
zones.

between 70 and 90% CaCO$_3$ depending upon the climate and several other variables (i.e., slope, length of exposure). Zones with a higher carbonate content than the weathering boundary form limestone, whereas zones with a lower carbonate content weather to marl. Thus, the carbonate content of the weathering boundary (C$_w$) and the carbonate content of the neutral value (C$_n$, which separates the sequence into dissolution and cementation zones) are usually not exactly equal (Fig. 14).

The box model data (Fig. 13) are not given for marl and limestone layers, but for dissolution and cementation zones. Moreover, <u>primary and present carbonate differences are not expressed for the amplitudes of the carbonate curves, but for mean integrated carbonate contents of the dissolution and cementation zones instead</u> (Fig. 14). Consequently, the factor of diagenetic enhancement is calculated between the mean primary and mean present carbonate differences in both zones. To calculate the amplitudes of carbonate oscillations, the mean differences for dissolution and cementation zones must be multiplied by a factor of 1.5 to 2.

2.4 Onset of Lithification

The hardening of calcareous ooze is a decisive factor in the diagenetic enhancement of bedding phenomena and rhythmicity (sections 4 and 5). In this study, an attempt is made to pinpoint the onset of lithification by using rock compaction data. Since mechanical compaction ceases as cementation begins (due to the development of a rigid framework at the grain contacts), the smallest amount of compaction in the middle of the limestone layers is, at first glance, a measure of the initiation of cementation (in terms of overburden and time). Thus, the mean minimum compaction in the middle of all limestone layers in a sequence is here referred to as "compaction at the onset of lithification", or simply "lithification compaction" (K$_1$), which is equivalent to the amount of mechanical compaction in the middle of the limestone layers. In simple terms, small amounts of compaction at the onset of lithification usually indicate early cementation accompanied by little mechanical compaction. On the other hand, intense compaction at the onset of lithification usually implies the opposite. In order to quantify this relationship, the compaction at the onset of lithification will be transformed into its corresponding porosity. This allows one to estimate the timing and the amount of overburden by using porosity-overburden formulas.

Transformation of the compaction at the onset of lithification (K_1) into the corresponding porosity (n_1) is accomplished by using the following equation:

$$n_1[vol\%] = \frac{n_o - K_1}{1 - 0.01 K_1} \quad , \tag{8}$$

where n_o is the porosity of decompaction (see section 2.3.5). Since numerous date on porosity reduction in carbonate sequences have already been collected (SCHLANGER & DOUGLAS, 1974; HAMILTON, 1976; SCHOLLE, 1977; LOCKRIDGE & SCHOLLE, 1978; SCHMOKER & HALLEY, 1982; KOPF, 1983), the amount of overburden and the time required until cementation begins can be estimated from the porosity at the onset of lithification.

HAMILTON (1976) has developed regression formulae for pelagic carbonates from the Deep Sea Drilling Project, where the density of sediment (ρ_h) and the amount of overburden (h, in km) is expressed in terms of the primary density of sediment (ρ_o) and of an empirically found density gradient (d, in $g/cm^3 m \times 10^{-4}$):

$$\rho_h = \rho_o + dh \quad , \tag{9}$$

where $\qquad d = 19.35 - 33.2h + 25.5h^2 \quad .$

In this study, ρ_o can be obtained from primary porosities as a result of decompaction (section 2.3.5) using a grain density (ρ_m) of $2.7 g/cm^3$ and a sea water density (ρ_w) of $1.05 g/cm^3$ (HAMILTON, 1976). Porosity with overburden (n_h) results in:

$$n_h[vol\%] = \frac{\rho_h - \rho_m}{\rho_w - \rho_m} \times 100 \qquad \text{(HAMILTON, 1976,} \tag{10}$$
$$\text{equation 1)}$$

Compaction with overburden can be obtained from equation 8 , where K_1 must be substituted by K and n_1 by n.

The development of porosity and compaction through time with increasing overburden was calculated as follows: Decompacted sedimentation rates for the studied sections and for the rock column above were ascertained with the help of absolute time tables (VAN

HINTE, 1976a,b; KENNEDY & ODIN, 1982). Then, compaction and porosity were calculated (eq. 8, 9, and 10) for a specific, small interval of overburden (e.g., overburden corresponding to a period of 10,000 years). This interval was shortened according to increasing compaction. In the second loop, porosity and compaction were calculated for a new value of overburden consisting of the shortened, first interval plus the original interval. Then, the original interval was again shortened due to the new amount of compaction and added to the shortened interval of the first run, etc. The method can be used until there is approximately 500m of overburden. After this point, equation 9 yields unrealistic results as a consequence of the regression method used by HAMILTON (1976).

3 QUANTIFICATION OF CARBONATE DIAGENESIS IN MARL-LIMESTONE ALTERNATIONS

In this section, the previously described methods are applied to several examples of marl-limestone alternations. Ten sections are arranged in a loose succession from slightly hardened, prediagenetic sequences to diagenetically modified, well-bedded alternations. Decompaction and carbonate mass balance calculations were performed in order to evaluate the cement content, the primary composition, and the timing of the carbonate redistribution. Data obtained directly from the sections studied, from laboratory analyses, and calculations (Table 2) are compiled in the Appendices II and III in RICKEN (1985b).

It is left to the reader to decide whether he wants detailed information on the methods used and results obtained or whether he prefers simply to read the conclusions at the end of this chapter (section 3.6). Readers who would like a short overview are referred to the figures illustrating the studied sections - each of which gives special insights into diagenetic bedding:

Porto Empedocle section (Fig. 15) is a prediagenetic alternation which is still in the phase of mechanical compaction and pore space reduction. The Pliocene foraminiferal ooze displays a primary cyclic bedding with low carbonate oscillations.

Rheine section (Fig. 18) clearly shows the behavior of porosity in lithified alternations.

Angles 1 section (Fig. 20) presents the typical shapes of the carbonate curves for clay-rich alternations.

Angles 2 section (Fig. 23) displays how the carbonate curves change when the total carbonate content increases; moreover, the behavior of minor elements which are contained in the carbonate fraction is shown.

Angles 3 section (Fig. 25) is an exciting section, since it represents a sequence generated from a primary alternation with easily detected large-scale fluctuations in carbonate content. In addition, Angles 3 is a typical example of how limestone layers should appear when their carbonate content is very high.

Logis du Pin section (Fig. 28). This is the best section for the demonstration of how compaction can be indirectly evaluated from selectively cemented burrows.

<u>Neuffen 1 section</u> (Fig. 32). This section clearly shows both the impact of weathering on the development of bedding in calcareous rocks and the different shapes of carbonate curves as reflections of low and high carbonate content.

Table 2: Field and laboratory analyses.

Section studied	Age	Carbonate cont. rock matrix	Carbonate cont. sel. cem. fossils, etc.	Compaction direct	Compaction indirect	Porosity	Elements in the carb. fraction (Mg, Sr, Fe, Mn)
Porto Empedocle Sicily	Pliocene	68	--	62	--	66	--
Rheine Lower Saxony Basin	Cenomanian	98	20	44	20	75	--
Angles 1 Vocontian Basin	Valanginian	110	7	67	7	36	--
Angles 2 Vocontian Basin	Hauterivan	97	29	73	29	85	68
Angles 3 Vocontian Basin	Barremian	68	17	20	17	15	--
Logis du Pin Vocontian Basin	Hauterivian	52	35	10	35	26	--
Neuffen 1 S. German Basin	Oxfordian	263	--	60	--	39	--
Neuffen 2 S. German Basin	Oxfordian	174	--	54	--	36	62, 122***
Gubbio 1 Umbrian Apennines	Paleocene Maastrichtian	127	15	68	15	34	--
Gubbio 2 Umbrian Apennines	Oligocene	144	47	44	47	24	143
Gubbio 3* Umbrian Apennines	Oligocene	33	27	6	27	--	--
Fossombrone** The Marches	Maastrichtian	77	3	31	3	60	--
Geisingen** S. German Basin	Oxfordian	55	--	--	--	--	--
Neuffen** S. German Basin	Kimmeridgian	93	--	--	--	--	--
Neuffen S. German Basin****	Oxfordian	87	--	--	--	--	--
Total amount of determinations		1546	200	539	200	496	385

*random spot sample, **event-alternation, ***only Mg and Ca, ****examination of stylolite.

<u>Neuffen 2 section</u> (Fig. 34) provides information on trace element behavior and demonstrates the shape of carbonate plots for carbonate-rich limestone layers.

<u>Gubbio 1 section</u> (Fig. 37). This section is a deep-water limestone – the layers are conspicuously thin and contain numerous stylolites which are oriented parallel to the bedding. It is an interesting example of how one can reconstruct primary sediments from compacted rock. In addition, Gubbio 1 spans the Cretaceous-Tertiary boundary as well.

<u>Gubbio 2 section</u> (Fig. 40) shows the abundance of minor elements in a stylolitic, deep-water alternation.

3.1   Uncemented Foraminiferal Marl, Pliocene, Sicily (PE)

This section serves as an example of an unlithified marl-chalk alternation with only weak carbonate fluctuations. It contains Pliocene foraminiferal marls and chalks called "Trubi" Marls, which outcrop on the southern coast of Sicily. The 90m thick Trubi Marls along with sandy transgressive sediments (BROLSMA, 1976) overlie Messinian gypsum sequences. Above the Trubi Marls is a 400m thick, clay-marl sequence that is intensely cyclic at points due to the repeated incursion of anoxic events. This entire Pliocene sequence is occasionally interbedded with debris flow deposits and turbidites which contain shallow water fossils (SPROVIERI, 1968; CITA, 1973; HEIMANN, 1977).

Most authors agree with HSÜ (1983) that the Mediterranean was a deep basin which became dessicated in the Miocene, and, after again becoming inundated, planktonic calcareous ooze (that is the existing Trubi Marl) was deposited. Micropaleontological evidence indicates that the Trubi Marls were deposited in water depths of 1000 to 2500m (CITA, 1972, 1973; BANDY, 1975; BENSON, 1975). However, HEIMANN (1977) hypothesizes a maximum water depth of only 500m (see conference volumes: DROOGER, 1973; BIJU-DUVAL et al., 1974; CATALANO et al., 1975; SAITO et al, 1975).

<u>Location</u>:   Coastal road between Porto Empedocle and Spoto, 500m westward of the intersection with the "Lido-Porto Empedocle" road.

<u>Section</u>:   The section (Fig. 15) is from the upper part of the Trubi Marls. True marl and limestone layers (and therefore weathered ledges) are not developed; however, slight carbonate variations do

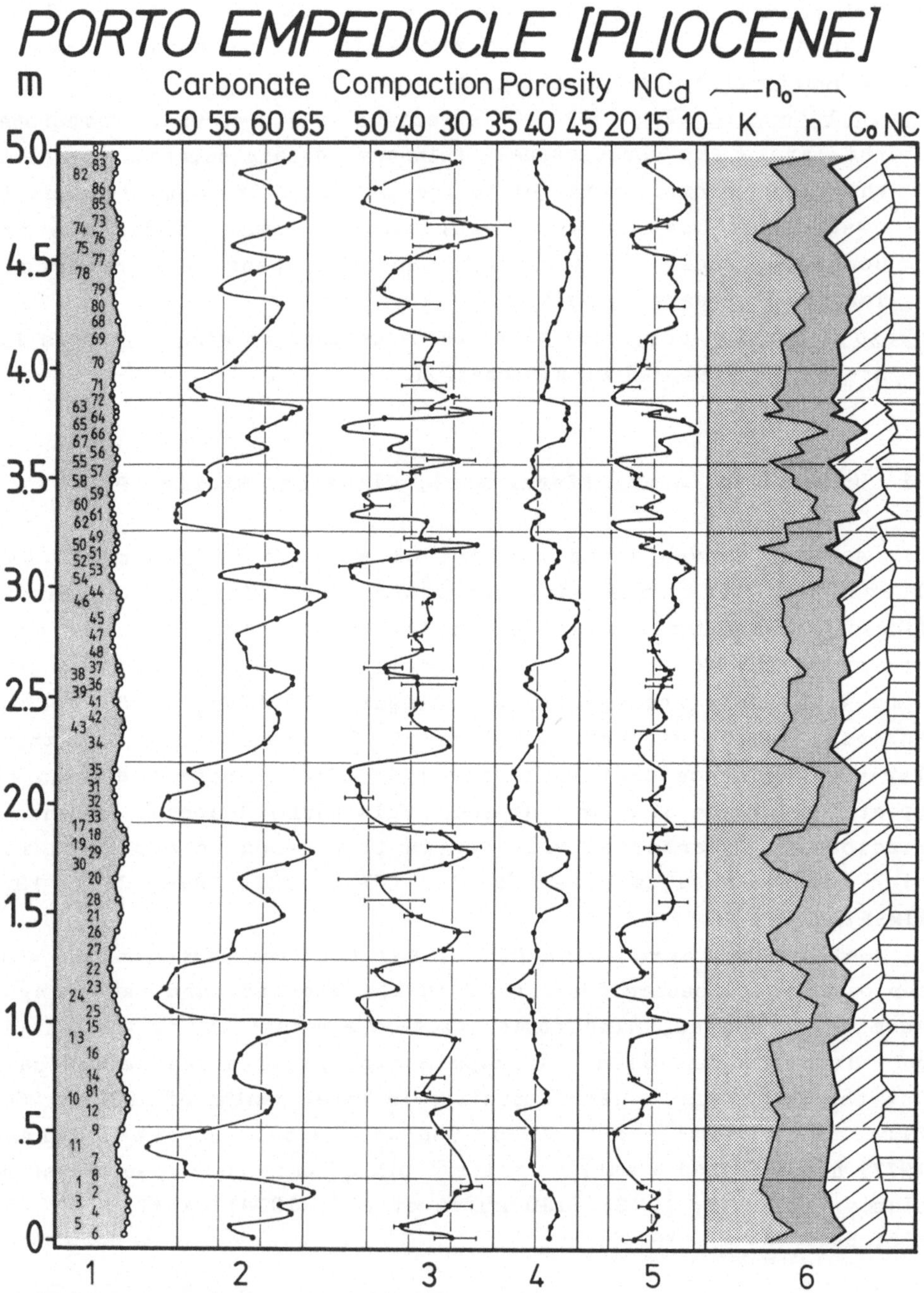

PORTO EMPEDOCLE [PLIOCENE]

Fig. 15 Uncemented foraminiferal marl, Pliocene, Porto Empedocle, Sicily (see Fig. 22A,B). Columns: 1) weathering profile and sample numbers, 2) carbonate content, 3) compaction (K), 4) porosity (n), 5) standardized noncarbonate fraction (NC_d), 6) calculated original sediment composition with the primary porosity (n_o, shaded area), original carbonate content (C_o, hatched area), and the standardized noncarbonate fraction (NC_d, striped area).

cause a weak undulation in the outcrop (Fig. 22A,B). The sequence is bioturbated with <u>Planolites</u>, <u>Chondrites</u>, and <u>Zoophycos</u>, and (due to 400m of overburden) is compacted by 35 to 50%. However, chambers of foraminifera often resisted the compaction of the matrix, although they have never been filled with cement. Carbonate oscillations have major periodicities of 1 to 1.5m with an amplitude of 15% $CaCO_3$. Superimposed upon these major oscillations are smaller carbonate variations which range from 0.2 to 0.3m and have fluctuation in $CaCO_3$ of up to 10% (Fig. 15).

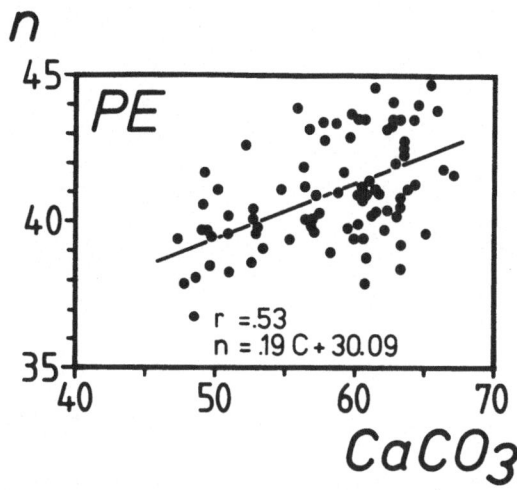

Fig. 16 Relationship between carbonate content and porosity. Unlithified foraminiferal marl, Pliocene, Sicily.

<u>Calculations and results</u>: Since a substantial amount of carbonate redistribution has not occurred, the curve of the absolute clay content (which is obtained from solving eq. 2) is in most parts inversely proportional to the $CaCO_3$ curve. Therefore, the nature of the original sediment can be determined by simply decompacting the existing section (Fig. 15). On the average, the primary sediment contained 63.4% carbonate, 36.6% clay, and a porosity of 64.0%. The decompaction calculation shows that the primary porosity was slightly higher in clay-rich beds as compared to that of the carbonate-rich beds, although in the section studied porosity is now lower in the marl beds (Fig. 16). This can be explained by a statistically higher amount of compaction in the layers with a relatively low $CaCO_3$ content (Fig. 17). The mean original differences in the decompaction porosity

(n_O) were only 1.5% between primary beds relatively richer and poorer in carbonate (for beds containing more than 57.5% carbonate: n_O = 63.5% based on 52 measurements; beds containing less than 57.5% carbonate: n_O = 65.0% from 26 measurements).

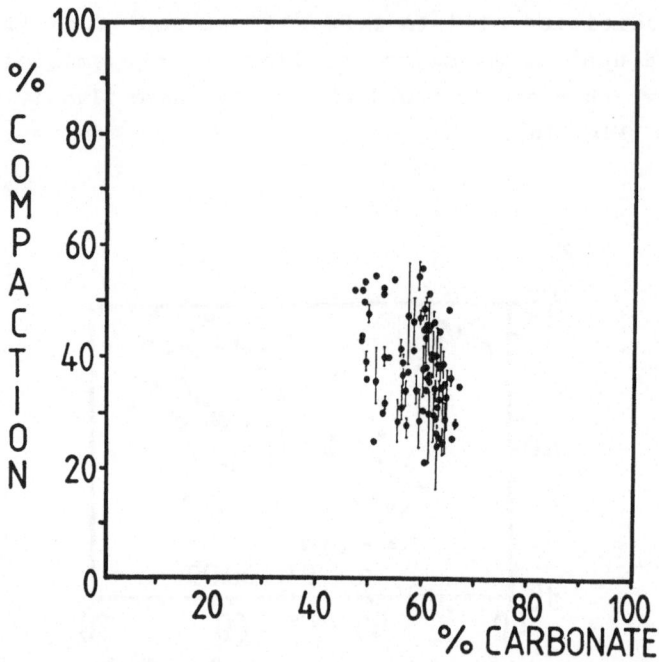

Fig. 17 Relationship between carbonate content and compaction. Unlithified foraminiferal marl, Pliocene, Sicily.

3.2 Marl-Chalk Alternation, Cenomanian, Lower Saxony Basin

Chalks and marls in the epicontinental Lower Saxony Basin in northern Germany have developed into marl-limestone or clay-marlstone alternations at certain locations (SCHNEIDER, 1964; ABU-MAARUF, 1975; ERNST et al., 1979). Although some alternations are difficult to correlate near salt domes (ABU MAARUF, 1975), ERNST et al. (1979) and FISCHER et al. (1985) suppose that they represent Milankovitch-like climatic bedding cycles.

    Location: Middle & Co. Quarry, Rheine, northwest Germany, Upper Cenomanian, below the Red Chalk layer.

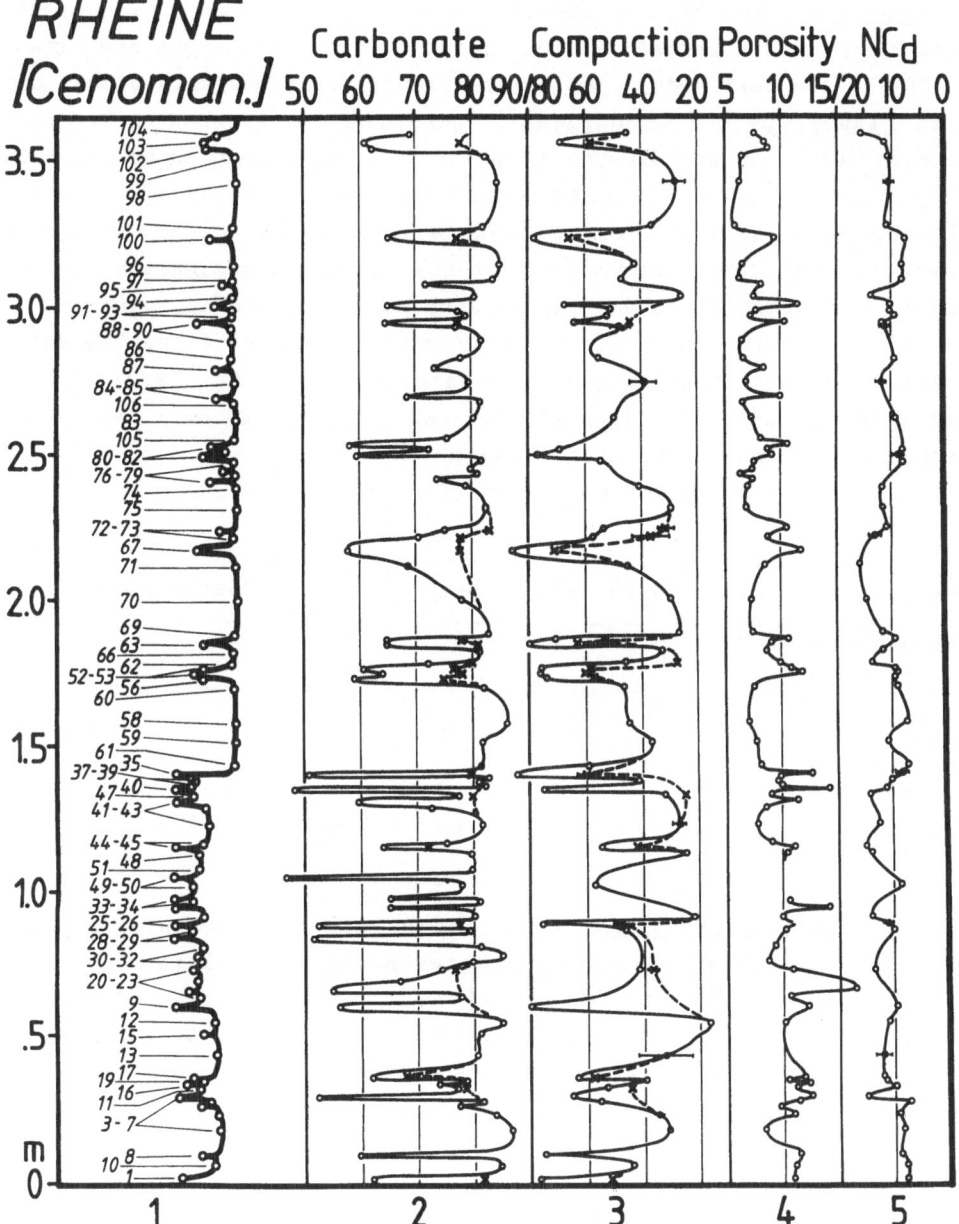

Fig. 18 Rheine marl-limestone alternation, Cenomanian, Lower Saxony Basin, Germany. Columns: 1) weathering profile and sample numbers, 2) carbonate content of selectively cemented burrows (x) and of the rock matrix (o), 3) compaction of selectively cemented burrows (x) and of the rock matrix (o), 4) rock porosity, 5) standardized noncarbonate fraction (NC_d).

Section: The Rheine section (Fig. 18) displays uneven bedding with 80 to 85% carbonate in the micritic limestone layers and with 50 to 60% in the marl beds. The alternations are intensively bioturbated with Planolites and Chondrites. These burrows have been selectively cemented early in the diagenetic history of the section (Fig. 22G). They have a higher carbonate content than the marl bed matrix, although they have a lower carbonate content and higher degree of compaction as compared to that of the neighboring limestone layers. Compaction of the limestone layers approaches 30% and approximately 70 to 80% in the marl beds. Porosity varies between 5 and 15% and is inversely proportional to the carbonate content (Fig. 19a); this is expressed by a relatively high marl bed porosity and by a decrease in carbonate content and porosity upsection. Sedimentary overburden of the section was approximately 1900m (THIERMANN, 1973).

Table 3: Mean standardized noncarbonate fraction (Rheine section).

Carbonate content of the rock (%)	<60	60-70	70-80	80
Mean standardized noncarbonate fraction (NC_d, vol %)	10.5	11.2	11.05	10.2
Standard deviation	2.5	2.9	2.2	2.1
Number of measurements	8	12	19	23

Calculations and results: The absolute noncarbonate fraction, calculated from eq. 2, varies between 7.5 and 15% and points to relatively turbulent depositional conditions. However, in the carbonate-compaction diagram (Fig. 19b), data are scattered along the theoretical curve calculated from the compaction law (where the mean absolute clay content is 10.7%). Since the mean absolute clay content

Fig. 19 Data from the Cenomanian Rheine section.
a: Relationship between porosity (n) and carbonate content (C).
b: Measurements and theoretical curve calculated from the compaction law using a mean absolute clay content of NC_d=10.7% and the formula for porosity used in Fig. 19a.
c,d: Carbonate mass balance showing histogram and box model.
e: Porosity (n), compaction (K), and time (t) in the sediment column, based on DSDP data (HAMILTON, 1976). Value shown represents compaction at the onset of lithification with a $\pm\sigma$ zone of scatter.

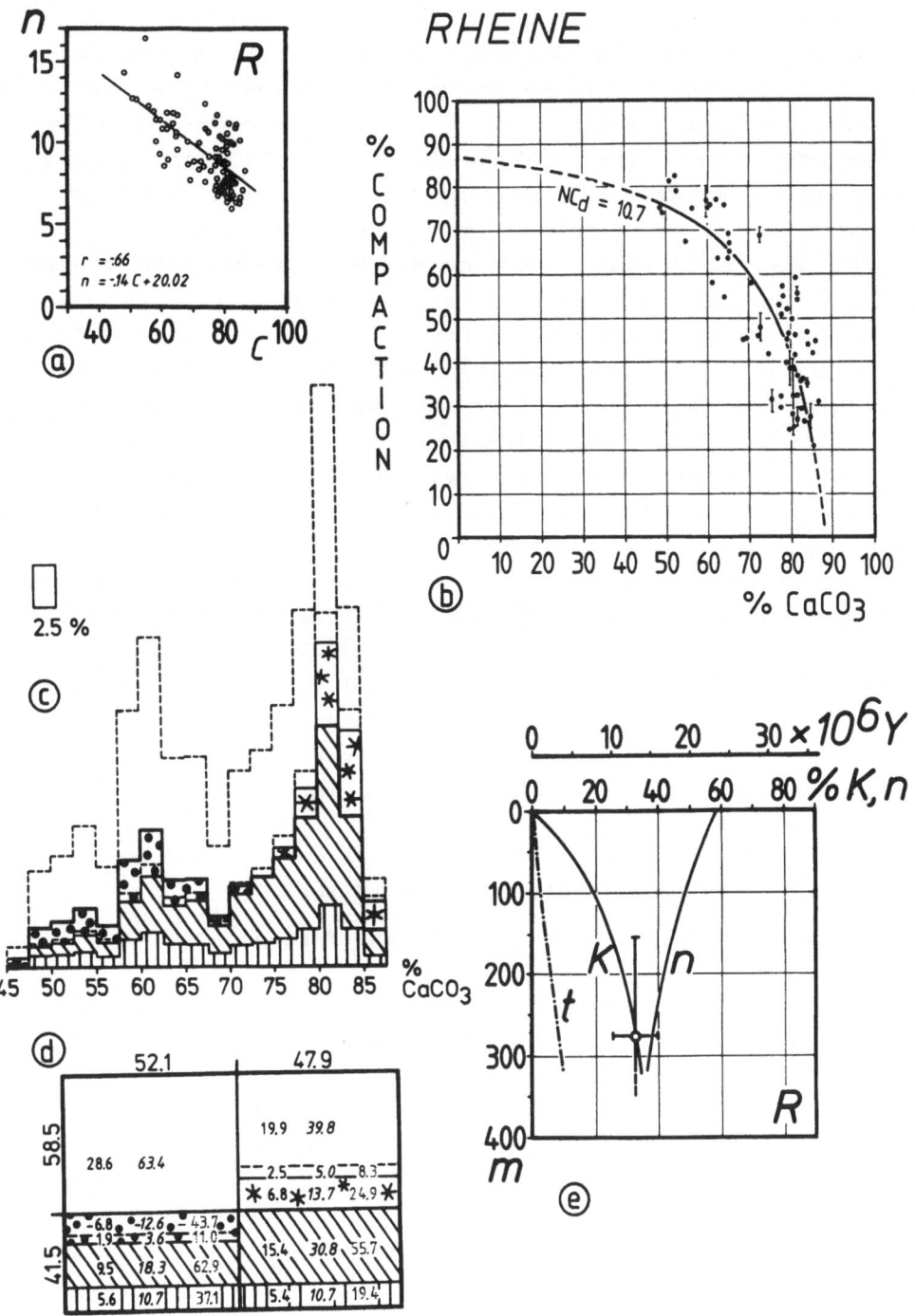

RHEINE

does not vary significantly with increasing carbonate content (Table 3), mass balance calculations were carried out using the mean absolute clay content. According to these calculations (Fig. 19c), the carbonate neutral value (at which carbonate is neither dissolved nor cemented) is about 74%. The carbonate in the limestone layers contains an average of 31% cement, whereas 41% of the original carbonate in the marl beds went into solution (Fig. 19d).

The mass balance calculation gives a mean decompaction porosity of 59% and a primary carbonate content of 74%. Originally, the mean carbonate content in the dissolution and cementation zones differed by a maximum of 2.8%, which resulted from the assumption of porosity differences of 5% in the original sediment (56 to 61% primary porosity with 72.6 to 75.5% $CaCO_3$, respectively). Now, however, the carbonate content in the dissolution and cementation zones differs by 17.7% (62.9 versus 80.6%, Fig. 19d). Thus, the minimum diagenetic enhancement is by a factor of 6.3.

The amount of compaction at the onset of cementation is defined by the minimum compaction in the middle of the limestone layers (see section 2.4). It has been calculated at 32.9% ± 15 and corresponds to a lithification porosity of 38%. According to the methods described in section 2.4, the phase of mechanical compaction lasted an average of 4 million years (decompacted sedimentation rates are 0.1m/1000y), and ended when 280m of overburden had been deposited.

3.3 Pelagic to Neritic Alternations of the
Vocontian Basin, Lower Cretaceous, France

The most conspicuous features of the Lower Cretaceous in the Vocontian Basin (French Maritime Alps) are the extremely well-bedded, pelagic to hemipelagic marl-limestone alternations (Fig. 22C) which interfinger with neritic sediments to the south (COTILLON, 1971; GEBHARD, 1983). Rhythmic alternations are interrupted by slumped horizons, and locally even by turbidites. COTILLON et al. (1980) have shown that several limestone layers can be traced across the tectonically shortened Vocontian Basin over a distance of 130km. Even sites which had been located over 2000km apart (Angles-Vergons section, Vocontian Basin, and sites in the Central Atlantic and in the Gulf of Mexico, sites 534 and 535/540) should display a similar bedding rhythmicity of 6000 to 26000 years (COTILLION & RIO, 1984). Limestone layers in the Vocontian Basin contain a significantly greater amount of radiolarians

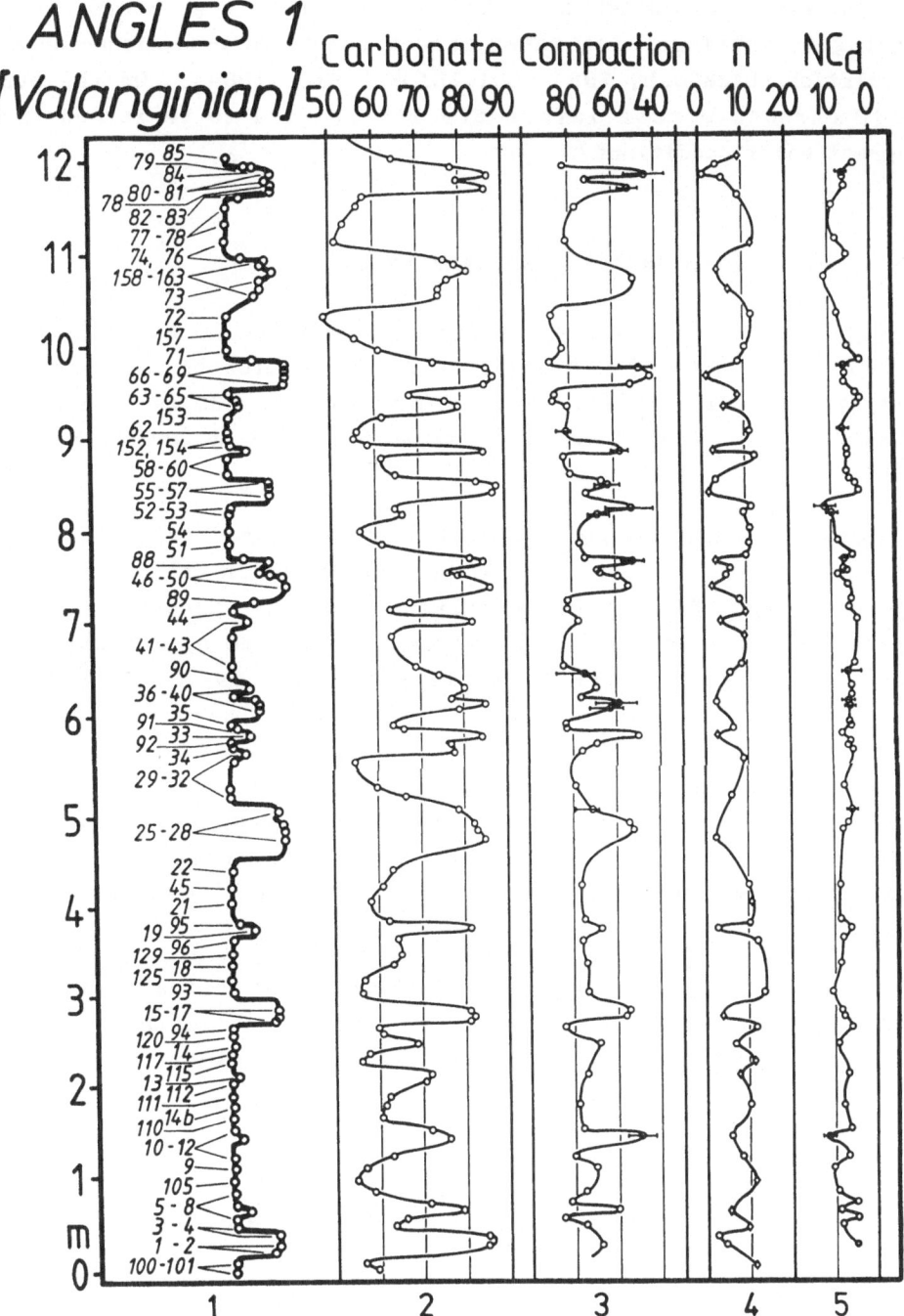

Fig. 20 Angles 1 marl-limestone alternation, Valanginian, Vocontian Basin, French Maritime Alps. Columns: 1) weathering profile and sample numbers, 2) carbonate content, 3) compaction, 4) porosity (n), 5) noncarbonate fraction, expressed as a percentage of the original sediment volume (NC_d).

than the marl beds (DARMEDRU et al., 1982). They are interpreted as forming in nutrient-rich water masses due to well-developed oceanic circulation (DARMEDRU, 1984; COTILLION & RIO, 1984). On the other hand, carbonate poor layers are interpreted to be formed under stagnant water conditions.

3.3.1 Marly Alternation, Valanginian (A1)

Location: Road cut along the road between Lac de Castillion and Angles, 15km north of Castellane. A 12m section of the Verrucosum zone, middle Valanginian, layer numbers 323 to 330 as assigned by DARMEDRU et al. (1982); see illustrated section by COTILLION et al. (1980).

Section: The section (Fig. 20) is comprised of one micritic marl-limestone alternation containing variations in carbonate content between 50 and 85% with periodicities of 0.8m. These variations represent relatively distinct maxima and minima on the carbonate curve. The grey, micritic marls and limestones are totally bioturbated; therefore, single, well-preserved burrows could not be detected. Thus, compaction was calculated from the living chambers of straight ammonites of the genus Anahamulina (preserved as steinkerns). Measurements indicate relatively high compaction in the middle of the limestone layers (40 to 60%) and about 80% in the marl beds. Sedimentary overburden was approximately 1250m (KERCKHOVE & ROUX, 1976).

Calculations and results: Absolute clay content, which was calculated from the compaction law (eq. 2) utilizing carbonate content, compaction, and porosity, decreases upsection somewhat

Fig. 21 Data from the Valanginian Angles 1 section.
a: Relationship between porosity and carbonate content.
b: Measurements and theoretical curve calculated from the compaction law using mean absolute clay contents of 6.0% and 7.4% and the formula for porosity used in Fig. 21a.
c,d: Carbonate mass balance showing histogram and box model.
e: Porosity (n), compaction (K), and time (t) in the sediment column, based on DSDP data (HAMILTON, 1976). Value shown represents compaction at the onset of lithification with a $\pm\sigma$ zone of scatter.

ANGLES 1

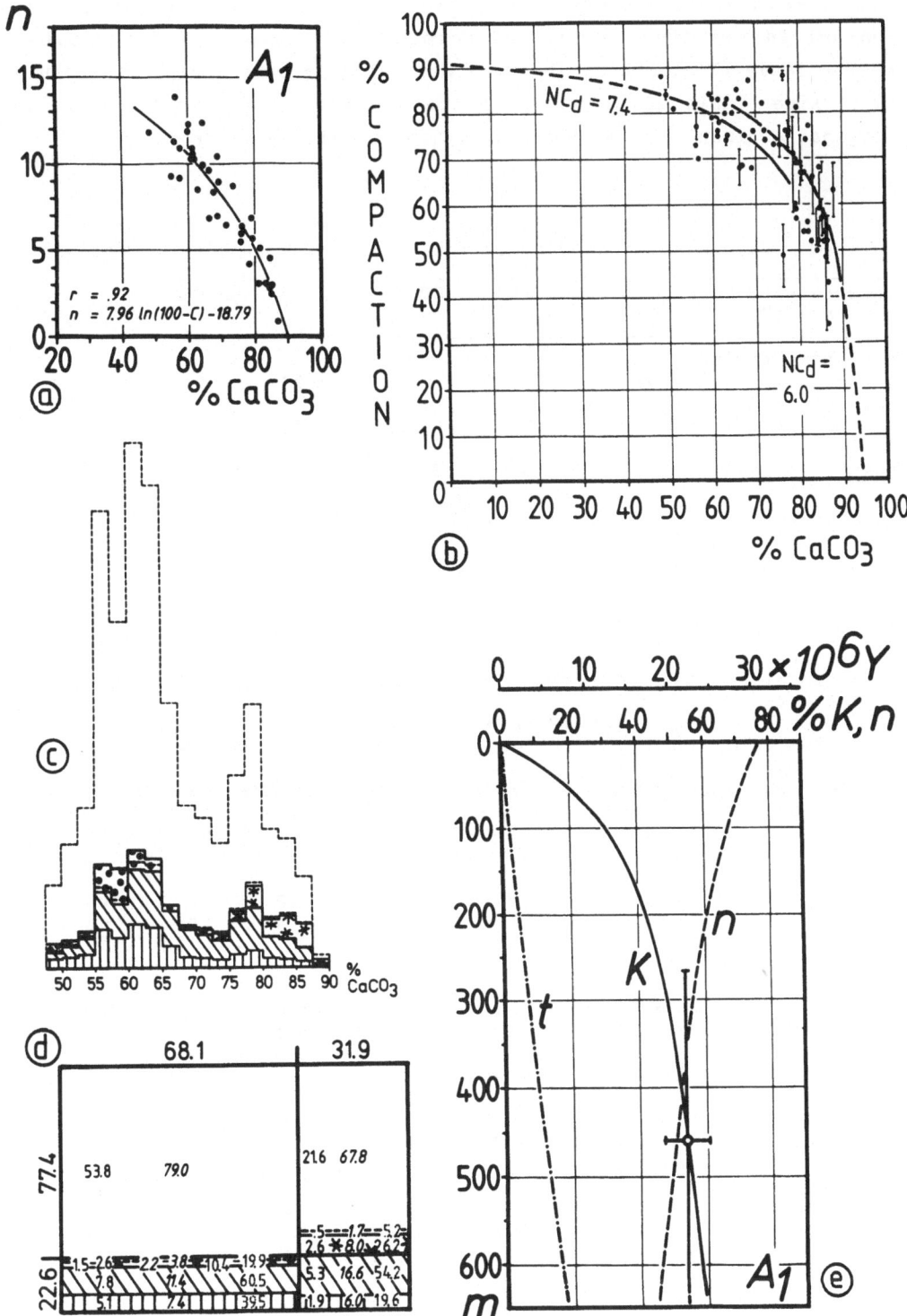

because the total carbonate content increases. This causes relatively
large scattering of the measured data around the theoretical curves
calculated from the compaction law, where the mean absolute clay
content in the dissolution and cementation zones is 7.4 and 6.0%,
respectively (Fig. 21b). The marl beds clearly have a higher absolute
clay value (Table 4), from which small, primary differences in the
original sediment can be derived without significant error.

Table 4: Mean standardized noncarbonate fraction (Angles 1 section).

Carbonate content of the rock (%)	<60	60-75	75-85	85-90
Mean standardized noncarbonate fraction (NC_d, vol %)	8.49	6.72	6.26	5.74
Standard deviation	1.61	1.95	2.43	1.10
Number of measurements	8	27	28	11

Fig. 22 A: Pliocene foraminiferal marl, Capo Rosello,
southern Sicily. Variations in carbonate content are easily
traced through the section, but no true bedding planes exist
(prediagenetic marl-chalk bedding cycles, marl-limestone
alternation Type 0).
B: Pliocene foraminiferal marls, section of Porto
Empedocle, southern Sicily. Again, no true bedding planes
exist.
C: Hauterivian part of the Angles section, Vocontian Basin,
French Maritime Alps. Marl and limestone layers display an
equal thickness. Limestone layers have a convex-shaped
distribution of their carbonate content (marl-limestone
alternation Type II).
D: Barremian Angles 3 section, Vocontian Basin, French
Maritime Alps. The sequence contains angular-shaped
carbonate curves and thin shale layers with a relatively
high organic carbon content (marl-limestone alternation Type
III).
E: Valanginian Angles 1 section, Vocontian Basin, French
Maritime Alps. Weathering points to irregular carbonate
distribution with several smaller limestone layers within
larger marl beds. Limestone layers have a predominantly
sinusoidal carbonate distribution (marl-limestone
alternation, Type I).
F,G: Selectively early cementation of the burrow system
(arthropod burrows), which suffered minor to intermediate
amounts of compaction prior to the onset of cementation.
F=Logis du Pin, Hauterive, French Maritime Alps; G=Rheine,
Cenomanian, Lower Saxony Basin, Germany.

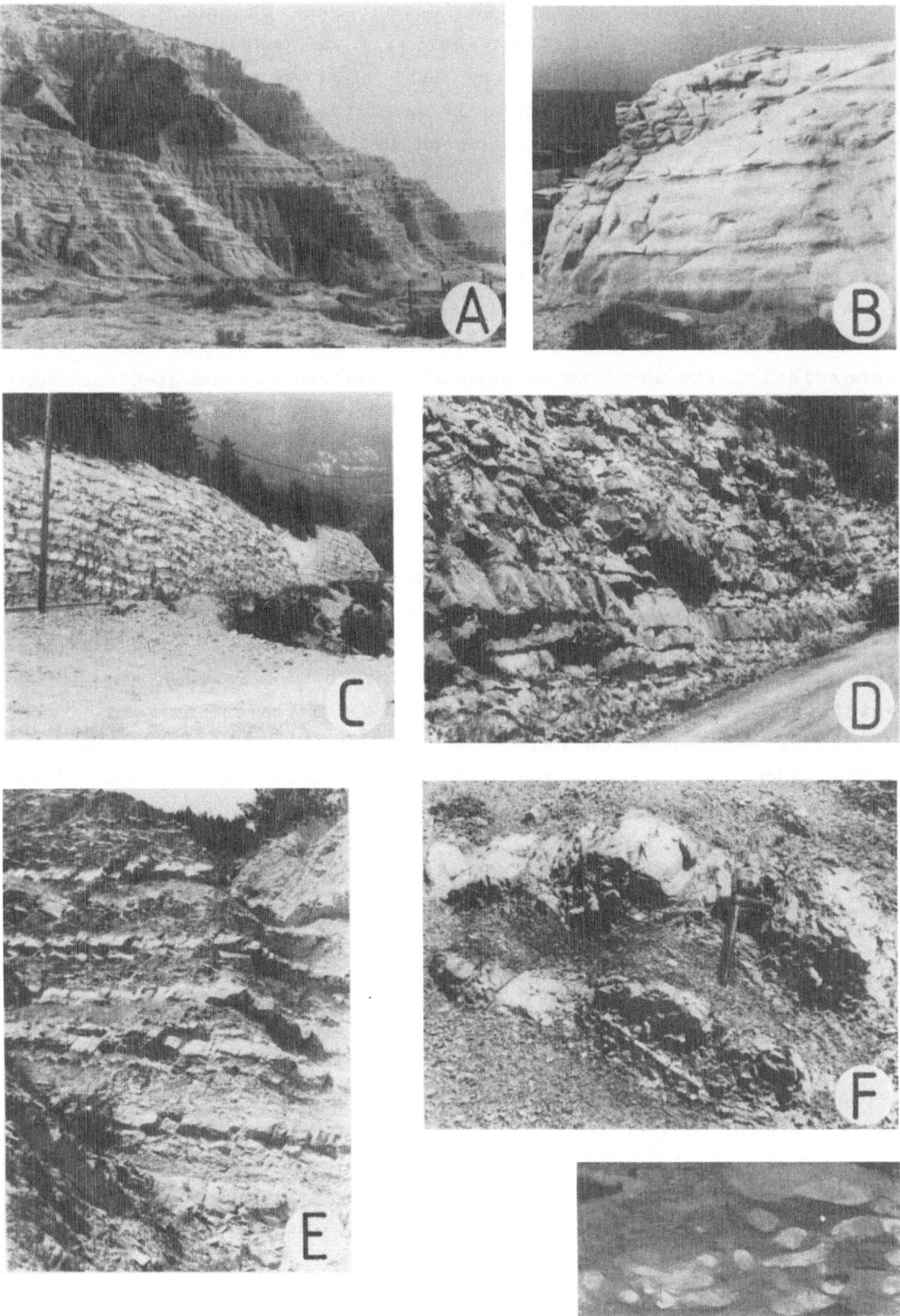

From the decompaction and the cement mass balance calculations (Fig. 21c,d), the mean primary porosity was computed as 77.4%. The statistical neutral carbonate value between dissolution and cementation zones is 70% (this is less than that at the weathering boundary); thus, the cemented zones are a little thicker than the limestone layers.

If one assumes porosity differences of 5% in the original sediment, primary mean carbonate differences between the dissolution and cementation zones would have a maximum of 13% (dissolution zones: n_O = 77.4+2.5%, C_O=63.2%; cementation zones: n_O=77.4-2.5%, C_O=76.1%). Mean diagenetic enhancement amounts to a factor of only 1.5. Accordingly, the absolute amounts of dissolved and cemented carbonate are low. Only 25% of the original carbonate content was dissolved from the marl beds. Nevertheless, the relative cement content in the limestone layers (33%) is not remarkably low, because the existing cemented zones are relatively small and have to incorporate the dissolved carbonate of the relatively thick dissolution zones (Fig. 21d). Mechanical compaction was high (45%) and porosity at the onset of the carbonate redistribution process was only about 51% which is in agreement with the relative small amount of diagenetically mobilized carbonate; thus, the corresponding mean overburden reached about 460m within an interval of 4.6 million years. Calculations were carried out using sedimentation rates of 0.16m/1000y (Valanginian) and 0.17m/1000y (Hauterivian), based on the thickness and time span of the decompacted sections.

3.3.2 Rhythmic Marl-Limestone Alternation, Hauterivian (A2)

Location: 1km west of Angles, along the road between Lac de Castillion and Angles (see limestone layers 2 to 16 in BUSNARDO, 1963).

Section: The section is nine meters thick and is from the Upper Hauterivian. Although the marl and limestone layers reveal similar thicknesses of about 30cm in the weathering profile (Fig. 22C), carbonate content is asymmetrically distributed with narrow minima and wide maxima (ranging between 50 and 90%) and exhibiting intervals of 0.6m (Fig. 23). Since burrows suitable for compaction measurements are lacking, compaction was evaluated from ammonites (Anahamulina and Crioceratites in steinkern preservation). Measurements indicate

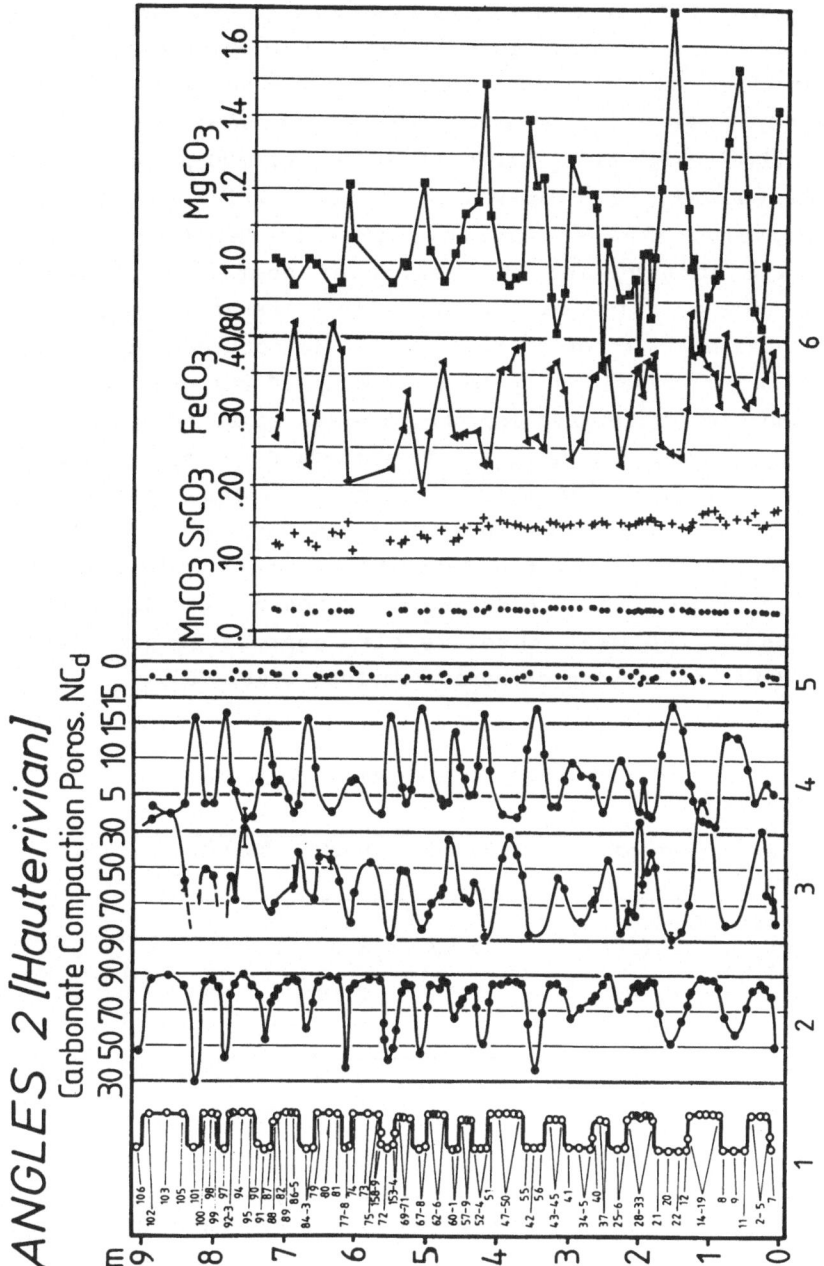

Fig. 23 Angles 2 marl-limestone alternation, Hauterivian, French Maritime Alps. Columns: 1) weathering profile and sample numbers, 2) carbonate content, 3) compaction, 4) porosity, 5) noncarbonate fraction, expressed as a percentage of the original sediment volume, 6) minor elements, expressed as a percentage of the total carbonate fraction.

compaction of 30 to 50% in the limestone layers and 80 to 90% in the marl beds. Rock porosity is inversely proportional to the carbonate content (Figs. 23, 24a) and varies between a few and 15%.

Minor elements contained in the carbonate (Fig. 23) have comparatively high $SrCO_3$ contents (0.15%, or 900ppm Sr), and low $MnCO_3$ values (0.03%, or 140ppm Mn). $FeCO_3$ and $MgCO_3$ (0.3% and 1%, respectively) show an inverse behavior depending on the total carbonate content and the degree of compaction. $MgCO_3$ becomes enriched by a factor of 1.3 to 2 within the marl beds, whereas $FeCO_3$ is depleted by a factor of 1.6 to 2.5. In contrast, Sr and Mn show no recognizable relationship to the amount of carbonate in the marl and limestone layers (further explanations are given in section 7).

Calculations and results: The standardized noncarbonate fraction (NC_d, eq. 2) varies between 4 and 8%. In the carbonate-compaction diagram (Fig. 24b), the measured data scatter around the theoretical curve, which is calculated from the compaction law (where the mean absolute clay content is 6.4% and the porosity formula is that used in Fig. 24a). The mean absolute clay content is not significantly affected by the varying carbonate content (Table 5).

Table 5: Mean standardized noncarbonate fraction (Angles 2 section).

Carbonate content of the rock (%)	30-70	70-80	80-90
Mean standardized noncarbonate fraction (NC_d, vol %)	5.32	6.52	6.61
Standard deviation	1.24	1.16	1.58
Number of measurements	8	15	38

Fig. 24 Data from the Hauterivian Angles 2 section.
a: Relationship between porosity (n) and carbonate content (C).
b: Measurements and theoretical curve calculated from the compaction law using a mean absolute clay content of 6.4% and the formula for porosity used in Fig. 24a.
c,d: Carbonate mass balance showing histogram and box model.
e: Porosity (n), compaction (K), and time (t) in the sediment column, based on DSDP data (HAMILTON, 1976). Value shown represents compaction at the onset of lithification with a $\pm\sigma$ zone of scatter.

ANGLES 2

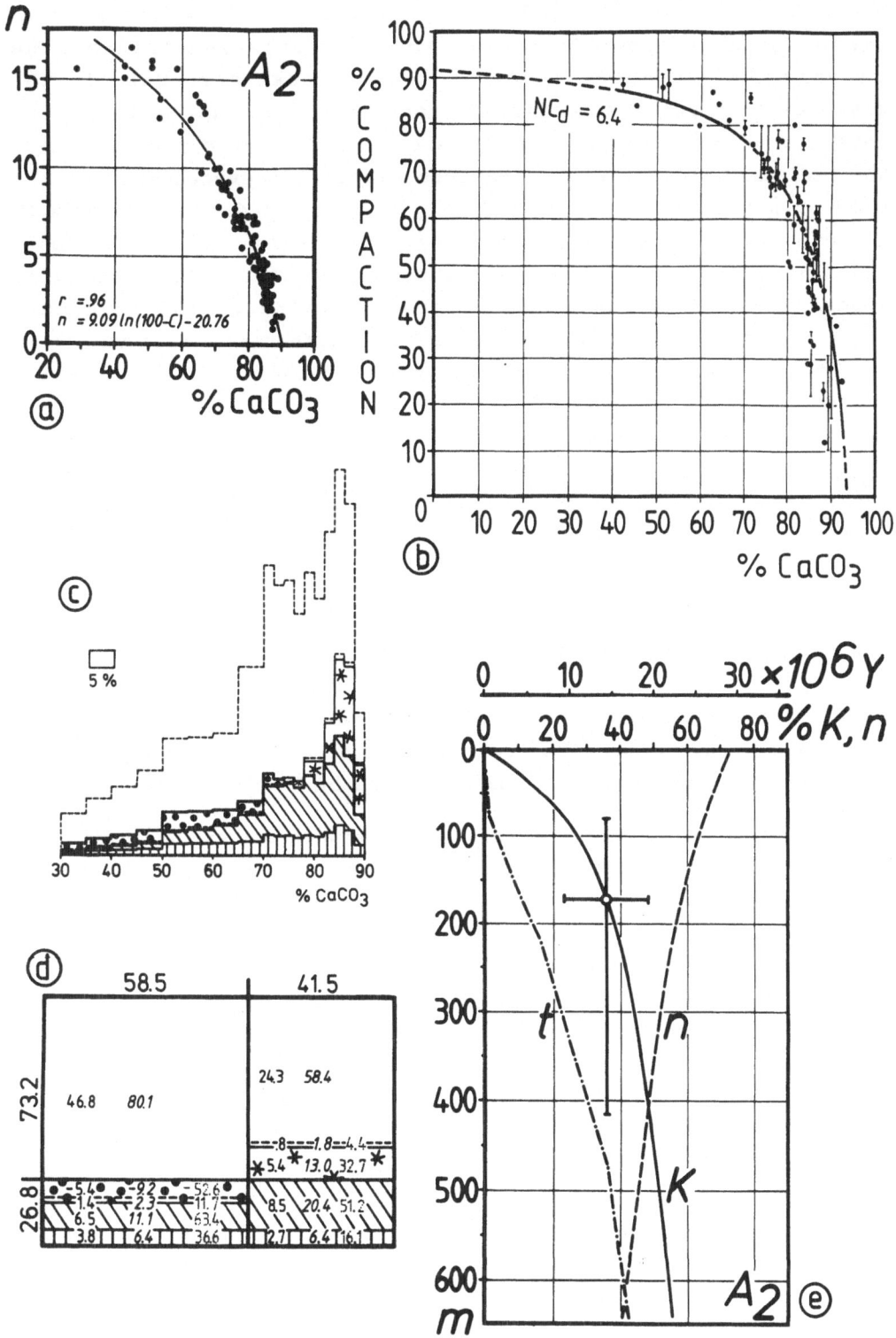

Decompaction and mass balance calculations (Fig. 24c,d) produce a mean primary porosity of 73.2% and a primary carbonate content of about 76%. Approximately 60% of the original sediment was affected by dissolution in order to supply the carbonate required by the cementation processes (45% of the original carbonate in the dissolution zones was mobilized, reprecipitated in the limestone layers, and now comprize 39% of the total carbonate of those layers). The redistribution created a maximum diagenetic enhancement of the mean original carbonate variations by a factor of 4.6. Assuming a 5% difference in the original porosity (see section 2.3.2), the mean primary carbonate differences between dissolution and cementation zones reach 4.5%. The neutral carbonate value of 76% is nearly equivalent to the carbonate content of the weathering boundary.

At a confidence level of 68% derived from the standard deviation of compaction at the onset of lithification, overburden was between 78 and 418m, with an average of 172m. The mean period of mechanical compaction lasted about 5 million years (decompacted sedimentation rates in m per 1000 years: Hauterivian, 0.17; Barremian, 0.04; Aptian, 0.06; Albian, 0.14).

3.3.3 Black Shale-Limestone Alternation, Barremian (A3)

Location: One km westward of Angles along the road between Lac de Castillion and Angles. The section from the Upper Barremian consists of limestone layers 144 to 150 as described by BUSNARDO (1963).

Section: The limestone layers are between 0.1 and 1 m thick and are extremely well-bedded. This bedding is especially pronounced due to the presence of bituminous marl to clay interbeds, most of which are only a few centimeters thick (COTILLION & RIO, 1984; see Fig. 22D). In contrast to the limestone layers, the bituminous marl layers are only slightly bioturbated by Chondrites, thus indicating anoxic or nearly anoxic conditions (BROMLEY et al., 1984). Flasery marl seams at the marl-limestone transition and pressure shadow structures around belemnite guards (see Fig. 58) within the marl beds point to intensive carbonate redistribution processes.

Compared to the previously described sections, these limestone layers (when plotted) have extremely angular carbonate curves (Fig. 25). This is typical for limestone layers with a high carbonate content and a relatively early onset of diagenesis. Carbonate content diminishes very slightly from the middle to the edges of the limestone layers, then decreases suddenly adjacent to the marl joints. Maximum carbonate differences are therefore between 5 and 97%. Compaction was evaluated using ammonites (mainly steinkerns of <u>Barremites</u> and <u>Crioceratites</u>) and <u>Chondrites</u>. The middle of the limestone layers is compacted by 10 to 40% of the original sediment thickness, whereas marl beds lost 80 to 90% of their original sediment volume. Maximum overburden was about 900m (KERCKHOVE & ROUX, 1976).

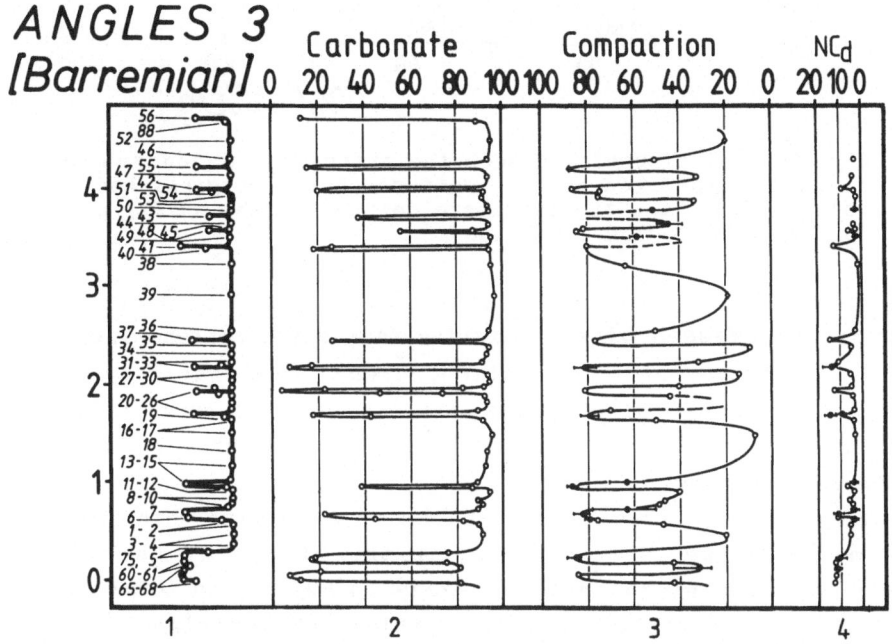

Fig. 25 Angles 3 section, Barremian, Vocontian Basin, French Maritime Alps. Columns: 1) weathering profile and sample numbers, 2) carbonate content, 3) compaction, 4) noncarbonate fraction, expressed as a percentage of the original sediment volume (NC_d).

<u>Calculations and results</u>: The standardized noncarbonate fraction (eq. 2) increases from 3 to 10% in the marl beds (Fig. 25) and clearly correlates with the existing fluctuations in carbonate content (Fig. 26). In the carbonate-compaction diagram (Fig. 27b), measured data lie along two separate curves calculated from the compaction law (where mean absolute clay contents in dissolution and cementation zones are 8.5 and 3.4%, respectively). Carbonate mass balance calculations result in primary sediment with mean carbonate contents ranging from 74 to 91% and an average porosity of 65% (Fig. 27c,d).

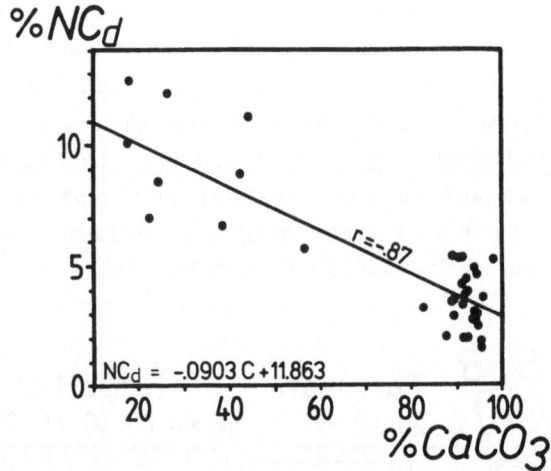

Fig. 26 Relationship between the standardized noncarbonate content (expressed as a percentage of the original sediment volume, NC_d) and the carbonate content in the existing rock in percent. Angles 3 section.

Primary thicknesses of the existing marl and limestone layers (areas of the dissolution and cementation zones) were nearly equal in the original sediment column. The carbonate neutral value is 90%, which is a somewhat higher percentage than the carbonate content of the weathering boundary.

The carbonate fraction of the limestone layers which is composed of cement averages 42%; however, in the middle of the limestone layers, the cement content increases to 60% of the total carbonate. About 70% of the original carbonate, which was contained in the marl beds, was dissolved and then reprecipitated in the limestone layers. Assuming primary porosity variations of 5% (see section 2.3.2), mean carbonate differences between the cemented and dissolved zones were

Fig. 27 Data from the Barremian Angles 3 section.
a: Relationship between porosity and carbonate content.
b: Measurements and theoretical curve calculated from the compaction law using mean absolute clay contents of 8.5 and 3.4% and the formula for porosity used in Fig. 27a.
c,d: Carbonate mass balance showing histogram and box model.
e: Porosity (n), compaction (K), and time (t) in the sediment column, based on DSDP data (HAMILTON, 1976). Value shown represents compaction at the onset of lithification with a $\pm\sigma$ zone of scatter.

ANGLES 3

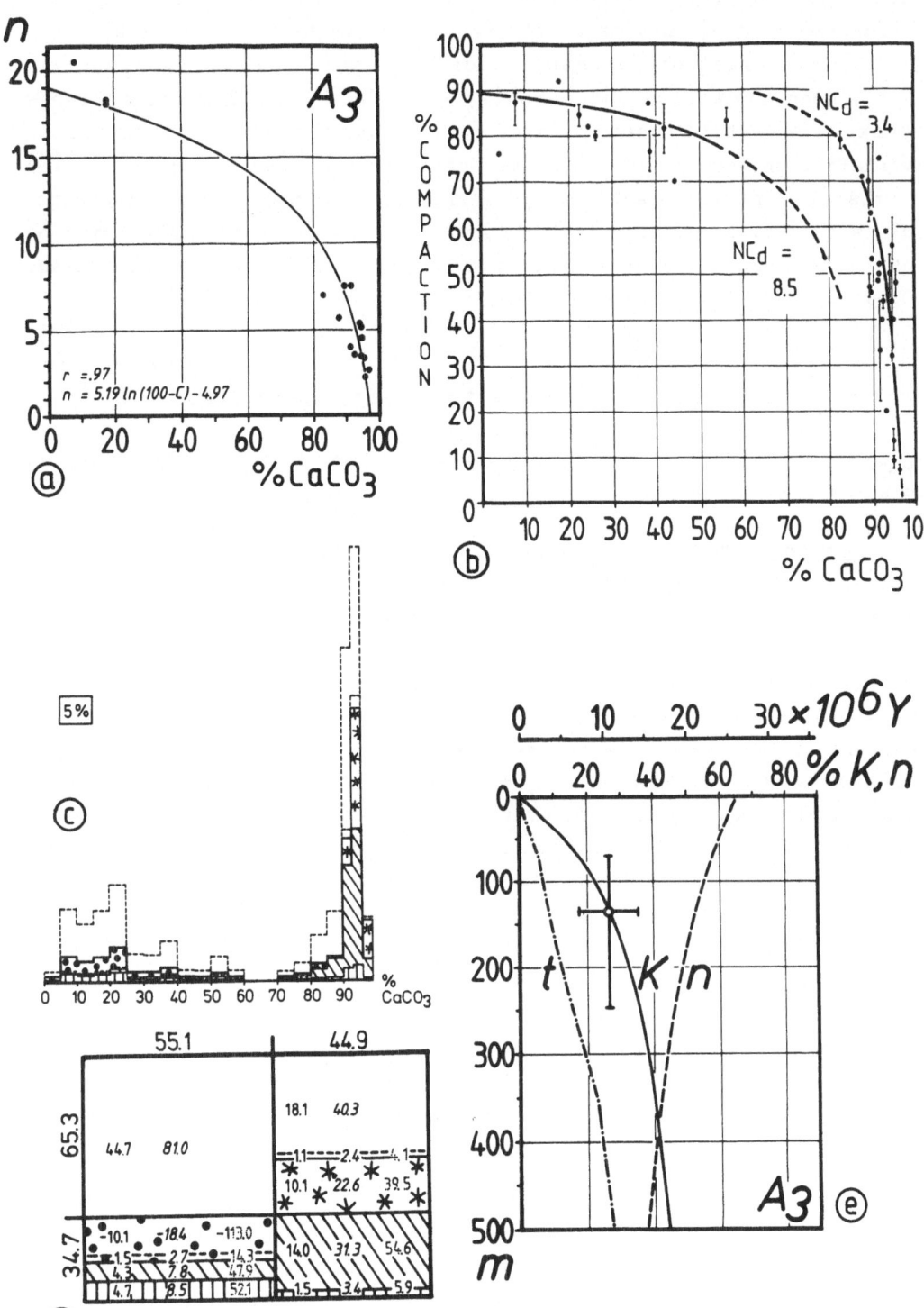

17.3% (73.6 and 90.9%, respectively), whereas present mean differences are 46.2% (47.9 and 94.1%, respectively); thus, minimum diagenetic enhancement of primary carbonate variations is by a factor of 2.7.

At the onset of intense carbonate redistribution (see section 2.4), mean compaction and porosity were 26.4 and 53%, respectively (Fig. 27e) which result from a mean overburden of 135m with a variation between 70 and 250m (according to the standard deviation for compaction at the onset of lithification). On the average, the phase of mechanical compaction lasted about 3.5 million years (calculated by using decompacted sedimentation rates in m/1000 years: Barremian, 0.04; Aptian, 0.06; Albian, 0.14).

3.3.4 Neritic Marl-Limestone Alternation, Hauterivian (L)

Location: Road cut and quarry 2km south of the crossroad "Logis du Pin" of R.N.85 and D.21.

Section: This upper Hauterivian section (Fig. 28) belongs to the neritic zone southeast of the Vocontian Basin and is composed of micritic, slightly silty marls and limestones, which are overlain by hardground-bearing, condensed Barremian carbonates (GEBHARD, 1983). The alternation is totally penetrated by bedding-parallel, selectively cemented arthropod burrows (Fig. 22F); they give the section a nodular character. The carbonate content of the burrow system, which varies from 90 to 95% in the middle of the limestone layers, is equivalent to that of the rock matrix. However, selectively cemented burrows within the marl layers usually contain less carbonate than the limestones but significantly more carbonate than the matrix of the marl layers, which consists of 70 to 75% $CaCO_3$. Compaction of the traces (20 to 70%) increases towards the middle of the marl layers; however, it is not as high as in the surrounding marl, which attained 75 to 80% compaction. Compaction in the marl layers was indirectly evaluated by considering the amounts of compaction and carbonate from the selectively cemented burrows (see section 2.1.2). Direct measurement of compaction was possible only on a few samples. All these were in close congruence with the indirectly evaluated compaction data.

Calculations and results: Standardized clay contents were calculated (eq. 2) from compaction, carbonate, and porosity data (Fig. 29a). They are between 5 and 10% and show only one major variation within the total section regardless of changes in lithology from marl to limestone. Since the mean absolute clay content fluctuates only slightly for different percentages of carbonate (Table 6), measured data scatter around the theoretical curve representing the relationship between carbonate content and compaction (Fig. 29b). The curve is calculated from the compaction law using a mean absolute clay content of 7.3% and the carbonate porosity regression curve in Fig. 29a.

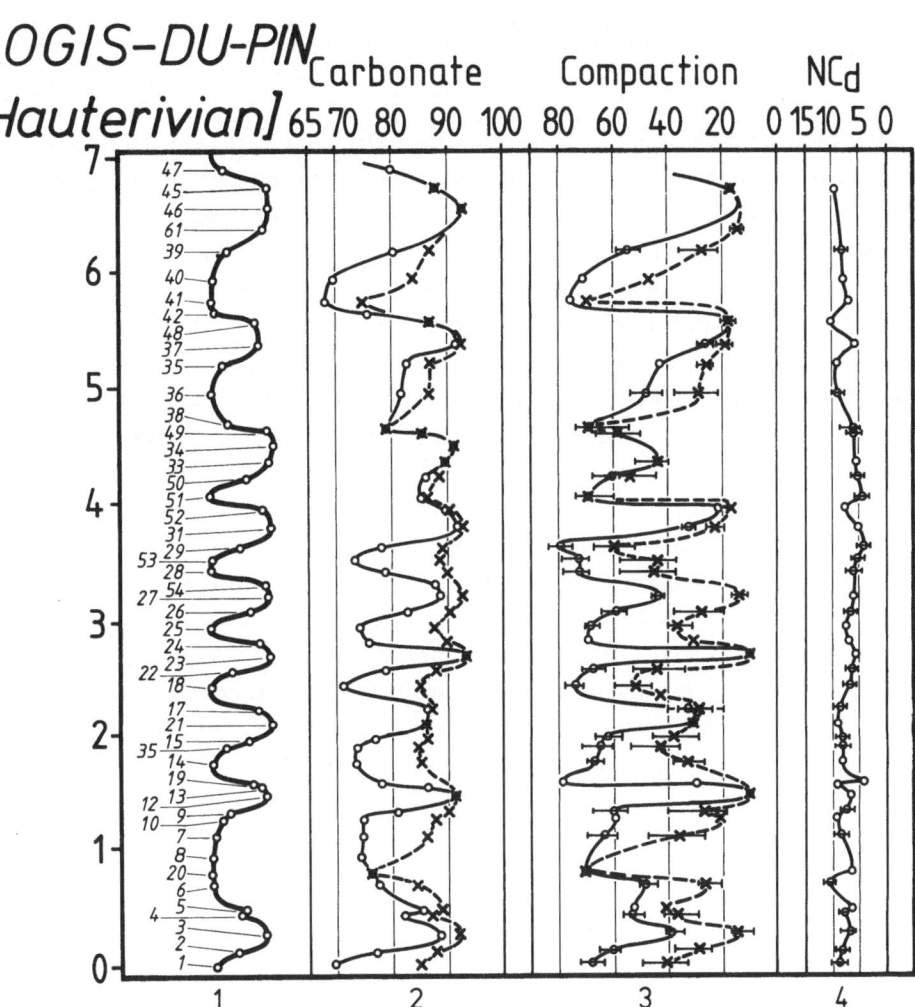

Fig. 28 Logis du Pin marl-limestone alternation, Hauterivian, Vocontian Basin, French Maritime Alps. Columns: 1) weathering profile and sample numbers, 2) carbonate content of mainly selectively cemented burrows (x) and rock matrix (o), 3) compaction of mainly selectively cemented burrows (x) and rock matrix (o), 4) noncarbonate content, expressed as a percentage of the original sediment volume (NC$_d$).

Table 6: Mean standardized noncarbonate fraction (Logis du Pin section).

Carbonate content of the rock (%)	70-75	75-80	80-85	85-90	90
Mean standardized noncarbonate fraction (NC_d, vol %)	7.78	6.63	7.62	7.59	6.22
Standard deviation	0.95	1.97	1.29	1.91	0.74
Number of measurements	10	10	6	14	5

Decompaction and carbonate mass balance calculations (Fig. 29c,d) were carried out for dissolution and cementation zones utilizing mean absolute clay contents of 7.3 and 7.2%, respectively. The decompaction porosity is relatively low, only 62%, which might be a result of the presence of silt. Its presence plays a critical role in reducing original porosity. Thirty-two percent of the primary carbonate in the dissolution zones was dissolved and then reprecipitated in what are now the limestone layers. Thus, a cement content of 36% was achieved (expressed as a percentage of the total carbonate content). The neutral carbonate value (81%) is lower than that for the marl-limestone weathering boundary (85%); thus the cemented portions of the alternation (or cementation zones) are thicker than the weathered portions, which form ledges in the outcrop (see Fig. 14). Mean original carbonate content fluctuated by approximately 2.5% (79.4% for dissolution and 81.9% for cementation zones), whereas the present mean carbonate differences between the two zones are 12.8% (74 and 86.8%); therefore, primary carbonate oscillations were enhanced by at least a factor of 5.1.

Fig. 29 Data from the Hauterivian Logis du Pin section.
a: Relationship between porosity (n) and carbonate content (C).
b: Measurements and theoretical curve calculated from the compaction law using a mean absolute clay content of 7.3% and the formula for porosity used in Fig. 29a.
c,d: Carbonate mass balance showing histogram and box model.
e: Porosity (n), compaction (K), and time (t) in the sediment column, based on DSDP data (HAMILTON, 1976). Value shown represents compaction at the onset of lithification with a $\pm\sigma$ zone of scatter.

LOGIS DU PIN

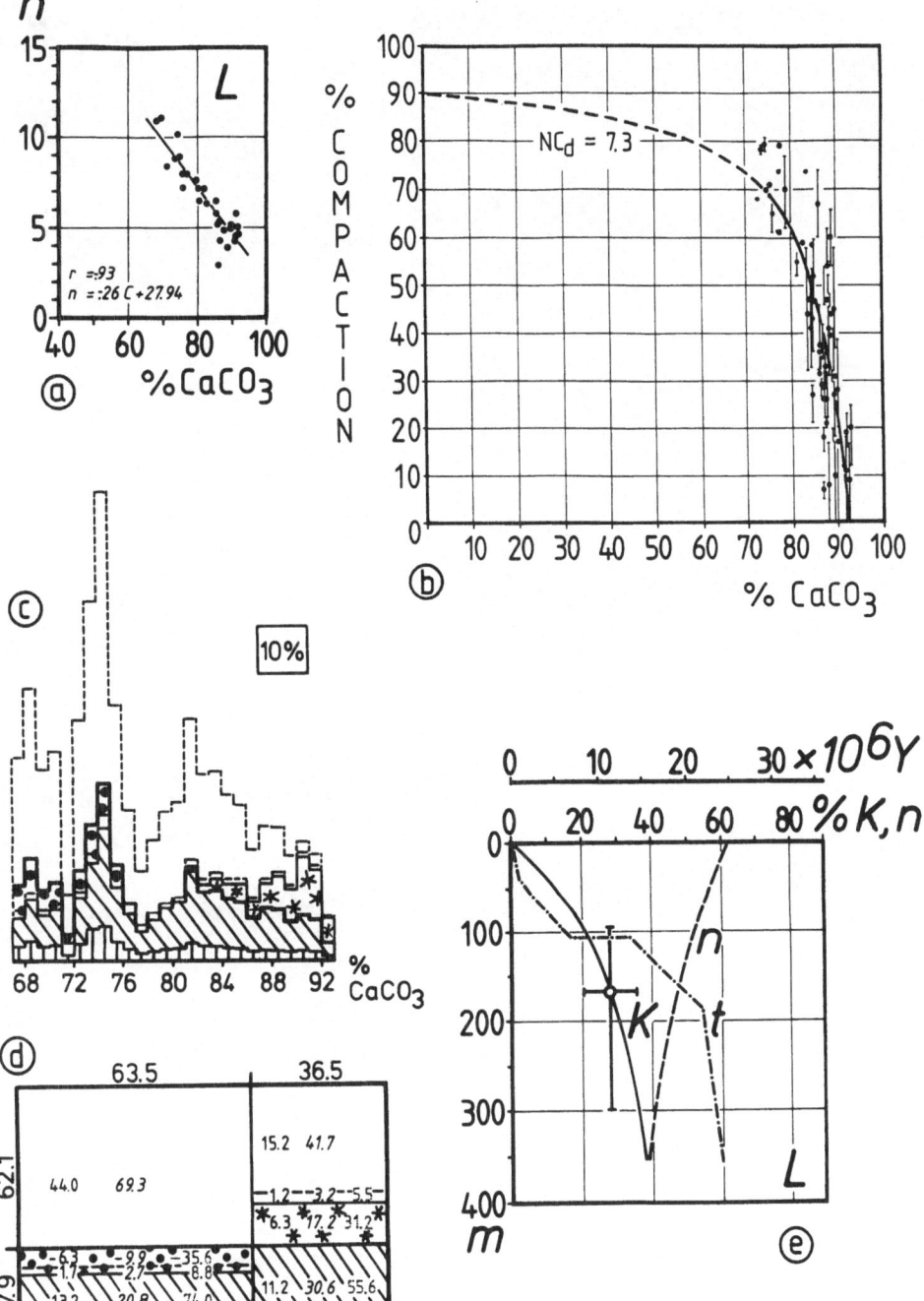

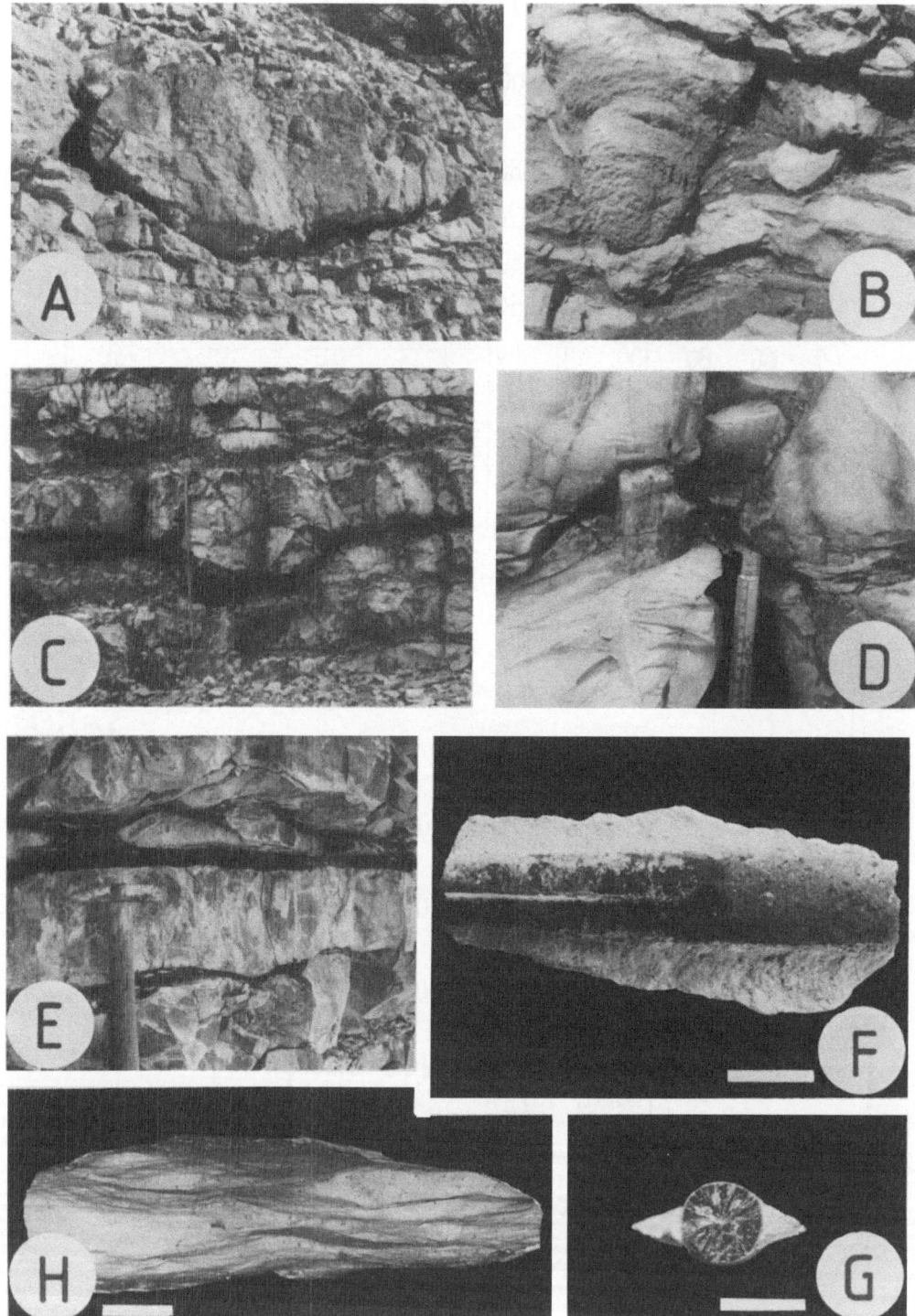

Compaction and porosity at the onset of cementation have been calculated as 27.9 and 47%, respectively. According to the methods presented in section 2.4, the resulting amount of overburden was between 95 and 300m with a mean of 170m. The phase of mechanical compaction had a long duration – 20 million years – because the Aptian is a period of sediment bypassing without netto-sedimentation (calculated decompacted sedimentation rates in m/1000 years: Barremian, 0.01; Aptian, 0.00; Albian 0.01; Cenomanian, 0.12).

3.4 Epicontinental Jurassic Alternations of Southern Germany

Since SEIBOLD's (1952) classic study (see section 5.6), the well-developed, marl-limestone alternations in southern Germany (Oxfordian and Kimmeridgian) have been thought to be the result of climatic bedding cycles. However, recent facies analyses (RICKEN, 1985a) indicate that the original bedding was partly formed by depositional events. However, the marl-limestone alternations now appear rather uniform because most of the depositional structures were destroyed by both intense bioturbation and diagenetic marl seam development (see section 8). The following observations confirm the principle of diagenetic bedding:

1. Marl layers of the Upper Oxfordian alternation display intensive flasering of microstylolitic dissolution seams (Fig. 30H, e.g., WANLESS, 1979); particularly in cored sections. These seams dissect the bioturbation structures and are partially associated with small-scale slickenslides as a result of differential compaction (RICKEN & HEMLEBEN, 1982). In addition, calcareous

Fig. 30 A,B,C: Diagenetic marl seams below and above slightly compacted algae-sponge bioherms (A,B: Gosheim, Middle Oxfordian) and skeletal channel fill (C: Neuffen, Upper Oxfordian; see Fig. 31).
D: Slickenslides due to compactional differences and development of diagenetic stylolites vertical to bedding, Neuffen, Upper Oxfordian (see Fig. 9).
E: Residual limestone layers within marl beds, Genkingen, Upper Oxfordian.
F,G: Typical $CaCO_3$ pressure shadow structure around a belemnite shell in a marl layer, Talheim, Upper Oxfordian (see Fig. 58). Scale is 1cm.
H: Typical microstylolitic, flasery marl seams as a result of pressure dissolution. Marl-limestone transition, Talheim, Upper Oxfordian. Scale is 1cm.

CARBONATE CONTENT

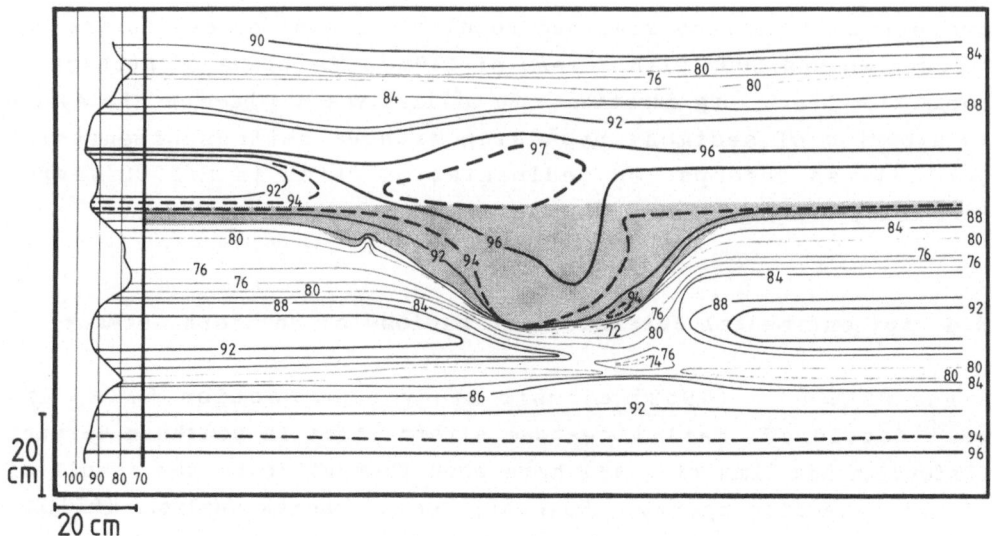

COMPACTION

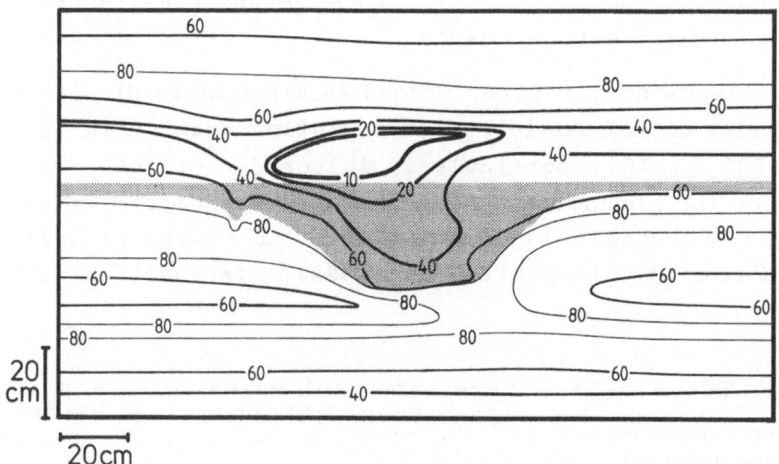

Fig. 31 CaCO₃ content and compaction of a bioclastic channel fill (shaded), Neuffen, Upper Oxfordian (see Fig. 30C). Below the weakly compacted channel fill, carbonate content decreases sharply as compaction increases. Data collected using a grid of points of 93 carbonate determinations. Compaction was calculated using the carbonate compaction law with an absolute clay content of $NC_d=-0.053C+7.8$.

pressure shadow structures around massive shells are present (see Figs. 30F,G; 58). Nevertheless, the weathered marl layers appear to be mostly homogenous.

2. Diagenetic marl layers formed above and below relatively little-compacted rock areas, such as several biostromes and coarse-grained channel fills (Figs. 30A,B,C; 31). The surrounding marl beds delivered part, but not all, of the carbonate necessary for the cementation of slightly compacted areas. Another source of cement are the dissolution seams within the biostromes.

3. Carbonate content fluctuates slightly when material from the marl bed was brought into the underlying limestone layer, except for in submarine channels and beds composed of lag deposits where the variations are more pronounced. This can be interpreted that the original sediment had smaller carbonate variations than the present, rhythmically bedded rock.

3.4.1 Middle Oxfordian Alternation (N1)

Location: Quarry near Neuffen, Swabian Alb, see SEIBOLD (1952) and RICKEN (1985a).

Section: The section (Fig. 32) comprises the upper part of the Middle Oxfordian, which is composed of blue-grey micrites and marls. The weathered section shows three intervals of closely spaced limestone layers, which are separated by 2m thick marl zones (Fig. 36A). In contrast to the appearance of the weathering profile, the section is composed of numerous carbonate oscillations (Fig. 32), which predominantly exhibit a major cyclicity of 5 to 8m. Small carbonate variations have narrow, sharp maxima on their carbonate curves, whereas larger carbonate variations mainly have curved, wide maxima. Below an average carbonate content of 85%, which is the weathering boundary (section 2.3.7), the total rock weathers to marl regardless of smaller carbonate variations. On the other hand, limestone layers develop when the carbonate content is above 85%. Compaction varies between 20% in the limestone layers and 80% in the marl beds. Compaction was evaluated using several trace fossils (Planolites, Chondrites) and ammonites (Glochiceras, Oppelia, Perisphinctes; in steinkern preservation).

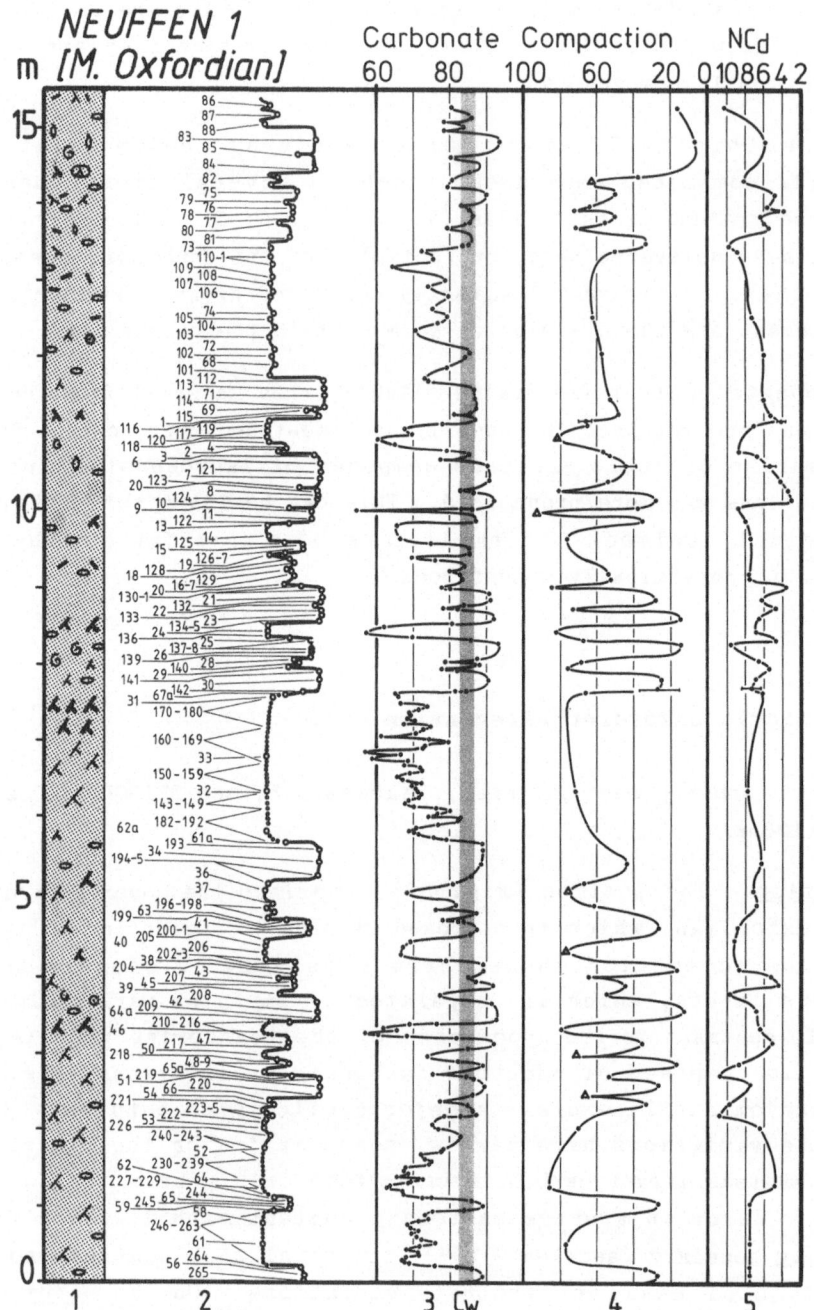

Fig. 32 Neuffen 1 marl-limestone alternation, Middle Oxfordian, southern Germany. Columns: 1) lithology, for explanation see Fig. 91, 2) weathering profile and sample numbers, 3) carbonate content and (shaded) the carbonate

<u>Calculations and results</u>: The absolute clay content, which was normalized to the original sediment volume (eq. 2), clearly varies (Fig. 32) and tends to be highest when the total carbonate content is lowest (Table 7). Decompaction and mass balance calculations (Fig. 33c,d) indicate a primary sediment with a mean porosity of 68% and a carbonate content of approximately 80%. Assuming a 5% porosity variation in the primary sediment, original mean carbonate differences amount to 5.4% (76.5% for dissolution and 81.9% for cementation zones); whereas present mean differences are 15.7% (71.1 and 86.8%, respectively). Therefore, the primary carbonate variations are diagenetically enhanced by a factor of 2.9. On the average, 32% of the original carbonate in the dissolution zones was released to form cement in the limestone layers with 37% of their total carbonate as cement. Statistically, the neutral carbonate value between dissolution and cementation zones (at 79%) is lower than at the weathering boundary; therefore, carbonate dissolution is restricted to the middle of the marl beds.

Table 7: Mean standardized noncarbonate fraction (Neuffen 1 section).

Carbonate content of the rock (%)	55-75	75-80	80-85	85-90	90-95
Mean standardized noncarbonate fraction (NC_d, vol %)	7.09	6.71	7.05	6.67	5.11
Standard deviation	1.47	2.08	2.14	1.62	1.26
Number of measurements	11	9	10	18	12

When cementation began, mechanical compaction was 27.4% thereby reducing the primary porosity to 55.5% (Fig. 33e). According to the methods described in section 2.4, the overburden at the end of mechanical compaction was between 50 and 350m (the average being 130m). The phase of mechanical compaction lasted about 1.2 million years. Decompacted sedimentation rates in m/1000 years: Oxfordian, 0.09; Kimmeridgian, 0.13.

◁content of the weathering boundary between marl and limestone (C_w), 4) measured compaction (●), interpolated by using eq. 4 and the mean absolute clay content (△), 5) noncarbonate fraction, expressed as a percentage of the original sediment volume (NC_d).

NEUFFEN 1

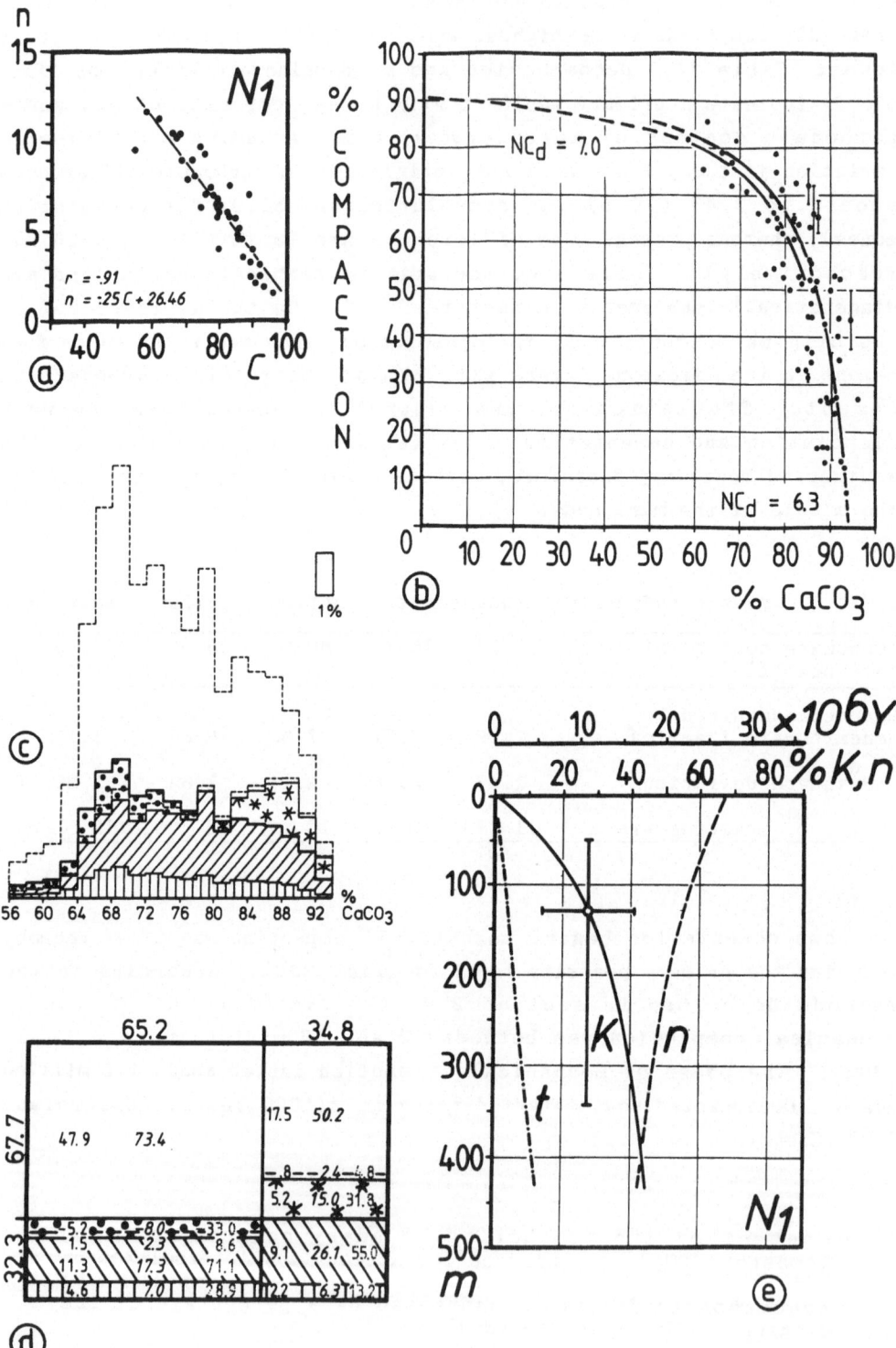

3.4.2 Upper Oxfordian Well-Bedded Limestones (N2)

Location: Quarry near Neuffen, Swabian Alb, section 10m below the
Oxfordian-Kimmeridgian boundary; see SEIBOLD (1952) and RICKEN
(1985a).

Section: The section (Fig. 34) is completely bioturbated and
represents maximum transgression of the Upper Oxfordian cyclothem (see
section 8). The weathering profile shows brick-like alternations of
20 to 30cm thick micritic limestone layers and 1 to 5cm thick marl
beds (Fig. 36B). The carbonate content varies between 60 and 97%; in
bedding-parallel stylolitic seams, it may drop to nearly 40% (see Fig.
9). When plotted, limestone layers produce angular carbonate curves.
These are characteristic of carbonate-rich alternations. Compaction,
calculated from the deformation of Planolites, is low only in the
middle of the limestone layers (5 to 30%), whereas it increases to 80%
in the marl layers. As shown later (section 7), carbonate-bound minor
elements are clearly dependent on fluctuations in the total carbonate
content and in the amount of compaction. In the carbonate fraction of
the marl beds (or dissolution zones) Mg and Fe are increased by
factors of 2 to 5 as compared to their concentration in the carbonate
of the limestone layers. On the other hand, Mn and Sr underwent only
slight enrichment in the residual carbonate of the marl beds.

Calculations and results: The calculated absolute clay content
(eq. 2) differs mainly within individual marl beds, which may be the
result of varied depositional events (see section 8). In the
carbonate-compaction diagram (Fig. 35b), measured data fall along two
theoretical curves (which are based on the carbonate compaction law
and mean absolute clay contents of 4.0 and 3.2% for the
dissolution-affected and cemented zones, respectively). The mean
absolute clay content tends to increase as carbonate content decreases
(Table 8).

Fig. 33 Data from the Middle Oxfordian Neuffen 1 section.
a: Relationship between porosity (n) and carbonate content
(C).
b: Measurements and theoretical curve calculated from the
compaction law using mean absolute clay contents of 7.0% and
6.3% and the formula for porosity used in Fig. 33a.
c,d: Carbonate mass balance showing histogram and box
model.
e: Porosity (n), compaction (K), and time (t) in the
sediment column, based on DSDP data (HAMILTON, 1976). Value
shown represents compaction at the onset of lithification
with a $\pm\sigma$ zone of scatter.

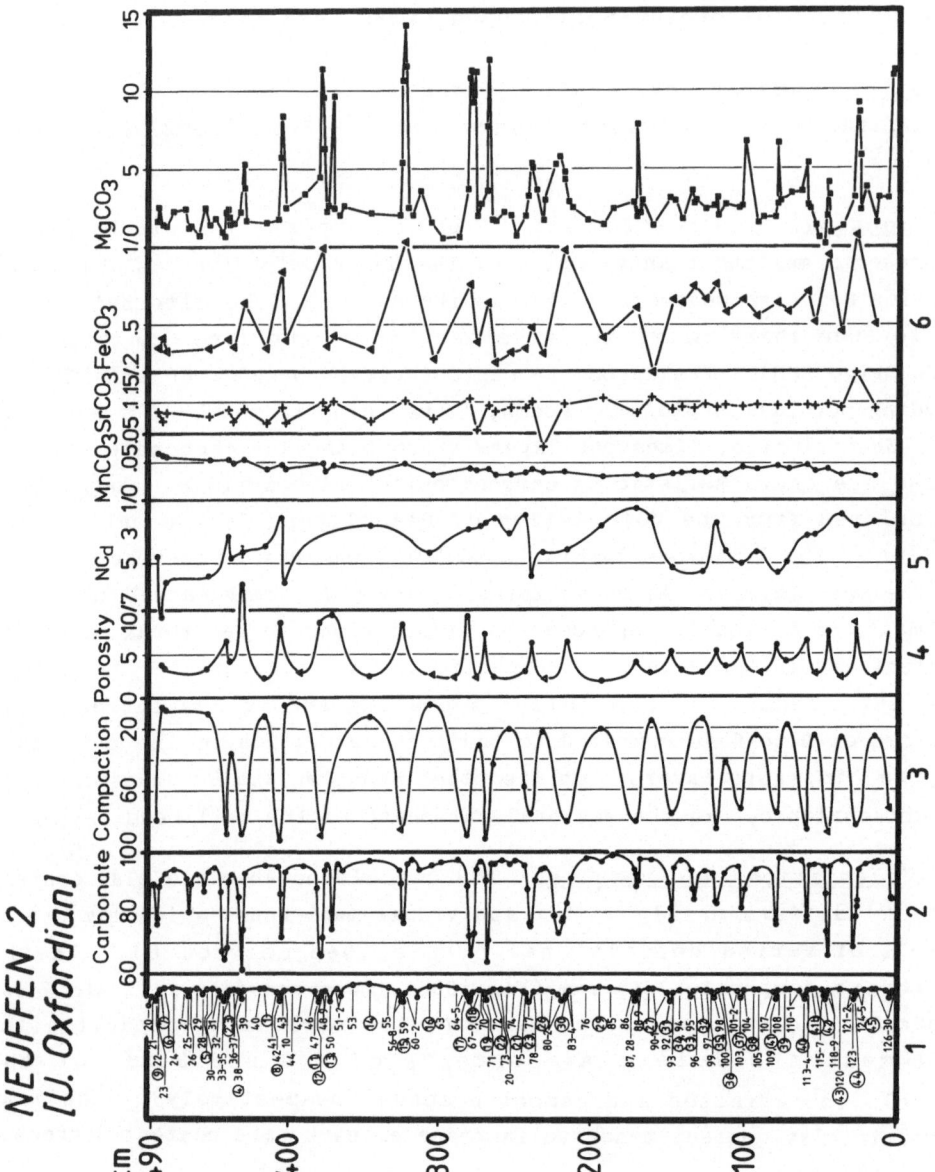

Fig. 34 Neuffen 2 marl-limestone alternation, Upper Oxfordian, southern Germany. Columns: 1) weathering profile and sample numbers, 2) carbonate content, 3) measured compaction (●) interpolated by using eq. 4 and the mean absolute clay content (Δ), 4) porosity, 5) noncarbonate fraction, expressed as a percentage of the original sediment volume (NC_d), 6) minor elements, expressed as a percentage of the total carbonate fraction.

Table 8: Mean standardized noncarbonate fraction
(Neuffen 2 section).

Carbonate content of the rock (%)	60-80	80-90	90-100
Mean standardized noncarbonate fraction (NC_d, vol %)	3.95	4.42	3.30
Standard deviation	1.29	2.50	1.34
Number of measurements	11	5	37

Assuming that the original porosity varied by 5%, carbonate mass
balance calculations yield mean primary carbonate contents of 87.6%
(for dissolution zones) and 91.4% (for cementation zones) and mean
original porosity of about 65%. Since mean primary carbonate
differences were 3.8% and the present rock has mean carbonate
fluctuations of 16.8% in dissolution and cementation zones, the
diagenetic enhancement of carbonate variations is by a factor of 4.4.
Due to the relatively small amount of compaction at the onset of
cementation, the large amount of pore space remaining in the limestone
layers was filled with carbonate cement (forming an average of 39% of
the total carbonate in the limestone layers). As a consequence of
this, a considerable amount (56%) of the original $CaCO_3$ in the
dissolution zones was released (Fig. 35d). The loss of volume in the
marl beds caused a large decrease in their thickness as compared to
the limestone layers, although originally the beds slightly richer and
poorer in carbonate had nearly the same thickness.

When carbonate redistribution began, an average mechanical
compaction of 19.3% reduced the original porosity to 57%; thus,
according to the methods described in section 2.4, a mean overburden
of 80m can be calculated with minimum and maximum values of 35 and
160m (this is due to the standard deviation in the measured compaction
at the onset of lithification). The phase of mechanical compaction
had a duration of about 0.8 million years. Decompacted sedimentation
rates in m/1000 years are: Oxfordian, 0.09; Kimmeridgian, 0.13.

3.5 Cretaceous to Tertiary Deep Water Limestones,

Umbrian Apennines, Italy

A complete, 500m thick sequence of Cretaceous to Tertiary deep water
sediments, predominantly composed of limestones, is exposed in two
nearby sections near Gubbio, Italy (in the Contessa valley and the

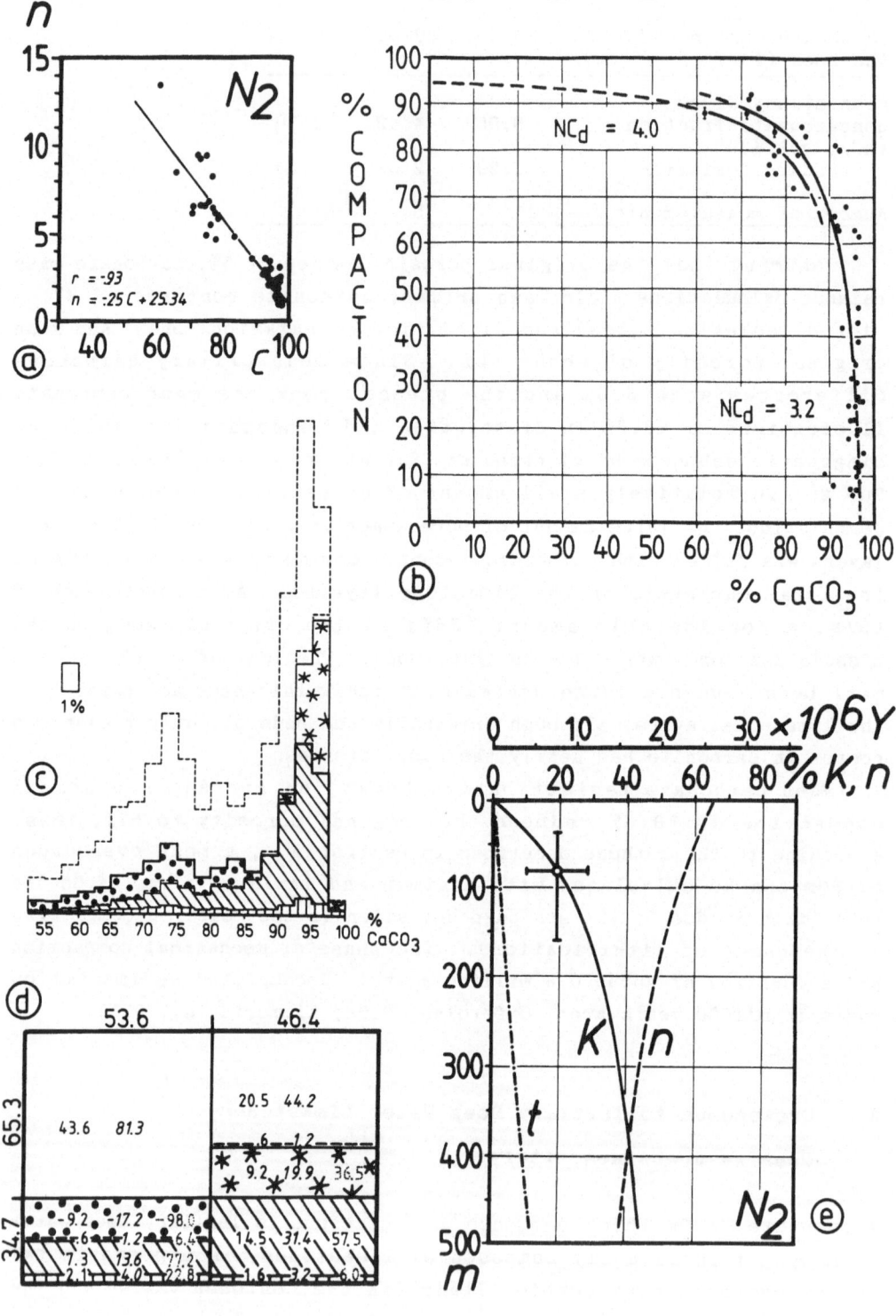

NEUFFEN 2

Bottacione gorge). The sequence was previously studied in detail in terms of its geochemistry and facies (ARTHUR, 1977; 1979; ARTHUR & FISCHER, 1977; ALVAREZ et al., 1980; 1981), biostratigraphy (LUTERBACHER et al., 1962; PREMOLI SILVA, 1977), and paleomagnetism (LOWRIE et al., 1977; ROGGENTHEN et al., 1977).

Other than the Albian shales and black shales, foraminiferal limestones of the Scaglia Bianca (Upper Albian to Lower Turonian) and Scaglia Rossa (Middle Turonian to Eocene) dominate the sequence. Limestones are free of macrofossils, and water depth during deposition is estimated to have been roughly 1000m (ARTHUR, 1979). In the Oligocene, limestones of the Scaglia Rossa grade into slightly silty, pelagic marls (Scaglia Cinerea), which are overlain by a 2.5km thick sequence composed of turbidite sandstones and shales (BORTOLOTTI et al., 1970). The bedding of the Scaglia resembles solution cleavage: Limestone layers, 5 to 20cm thick, contain irregular, stylolitic dissolution seams (Fig. 36F,G,H). ARTHUR (1979) has distinguished first and second order dissolution planes, for which he has identified bedding rhythmicities of about 100,000 and 20,000 years, respectively (see also FISCHER et al., 1985).

3.5.1 Stylolitic Limestones from the Cretaceous-
Tertiary Boundary (G1)

Location: Road cuts near Gubbio, Italy. The Maastrichtian was sampled in the Bottacione gorge; the Paleocene was sampled in the Contessa valley.

Section: The Cretaceous-Tertiary boundary at Gubbio (Fig. 37) is sharp, and it is well-defined by a 1 to 2cm thick clay layer (or "boundary clay"). The boundary is also characterized by an abrupt

Fig. 35 Data from the Upper Oxfordian Neuffen 2 section.
a: Relationship between porosity (n) and carbonate content (C).
b: Measurements and theoretical curve calculated from the compaction law using mean absolute clay contents of 4.0% and 3.2% and the formula for porosity used in Fig. 35a.
c,d: Carbonate mass balance showing histogram and box model.
e: Porosity (n), compaction (K), and time (t) in the sediment column, based on DSDP data (HAMILTON, 1976). Value shown represents compaction at the onset of lithification with a $\pm\sigma$ zone of scatter.

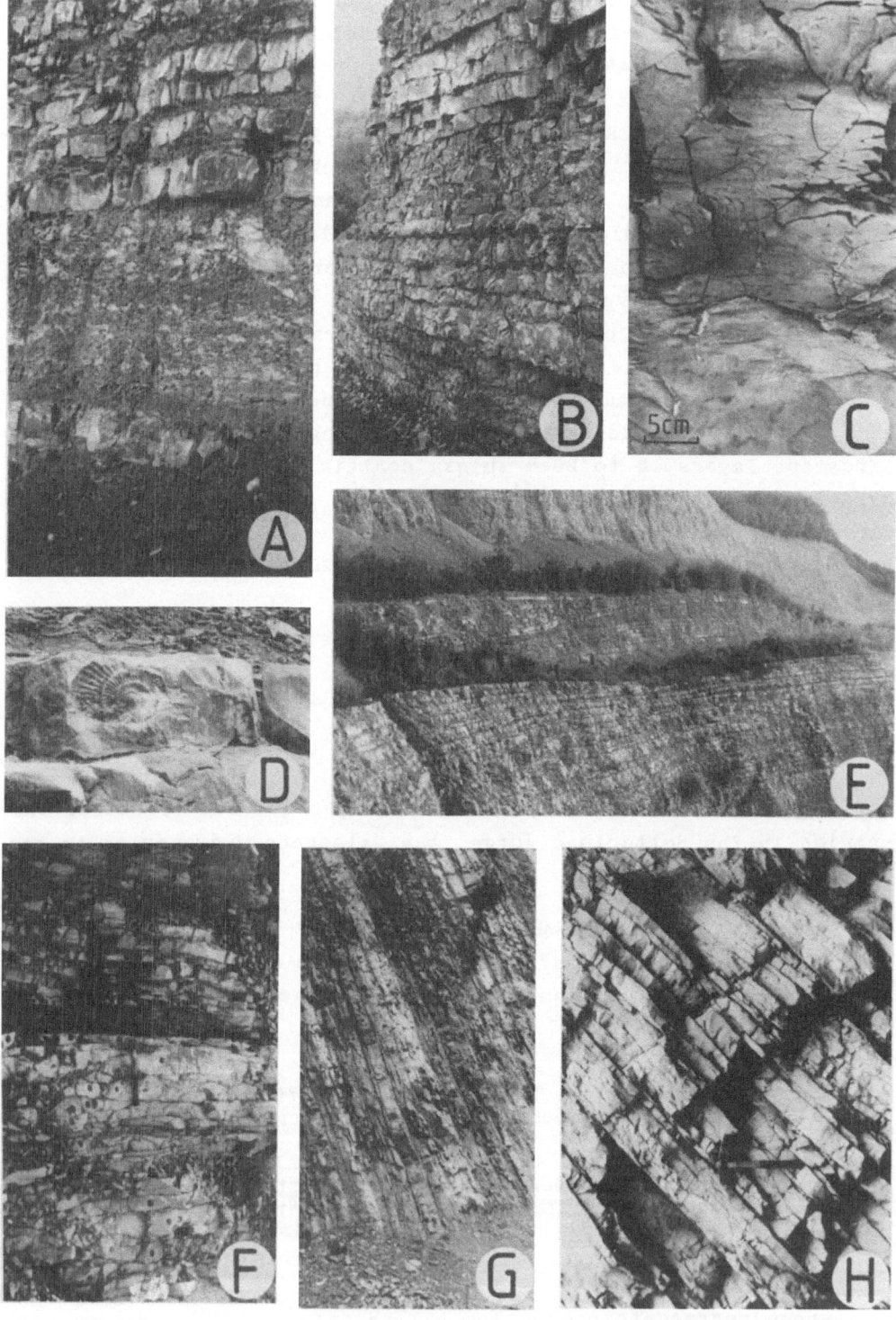

change in the foraminiferal assemblages (LUTERBACHER & PREMOLI SILVA, 1962). Since high iridium contents are found in the boundary clay of this and similar sections (ALVAREZ et al., 1980), the Cretaceous-Tertiary boundary is one of the most discussed and disputed stratigraphic events. Explanations of this event include extraterrestrial impact (e.g., ALVAREZ et al., 1980; 1982; 1984; O'KEEFE & AHRENS, 1982) and volcanic and environmental processes (e.g., KENT, 1981; KEITH, 1982; McLEAN, 1982; RAMPINO, 1982; EKDALE & BROMLEY, 1984).

The two meter thick interval studied contains numerous bedding-parallel stylolitic planes, in which the carbonate content (approximately 90% in the limestones) decreases to values ranging from 20 to 70% (Fig. 37). In the limestone layers (5 to 10cm thick), compaction could be directly evaluated by measuring the deformation of _Planolites_ and _Teichichnus_. However, compaction measurements in the stylolitic bedding planes usually were impossible.

Lower Tertiary rocks have completely cemented limestone layers with strongly reduced porosity and marl beds with much higher porosities; whereas porosity in the limestones and marls from the uppermost Cretaceous remains more or less constant at 3.5% (Figs. 37; 38a,b). Porosity data provide no evidence for early hardground

Fig. 36 A: Thick marl bed in the Neuffen 1 marl-limestone alternation, Middle Oxfordian. The marl bed contains small carbonate variations. However, the carbonate content of the variations is below the critical carbonate value of the marl-limestone weathering boundary, therefore the variations weather entirely to marl.
B: Marl-limestone alternation with high mean carbonate content, angular carbonate curves, and thin marl layers (Type III). Upper Oxfordian, Pfullingen, South Germany.
C: Compactional deformation of the bioturbation structure in a marl layer. Upper Oxfordian, Neuffen, southern Germany.
D: Weak compaction in limestone layers. Slightly deformed ammonite oriented vertical to bedding. Middle Oxfordian, Schlatt, South Germany.
E: Marl-limestone alternation in an ancient submarine channel system which shift laterally to the northeast. Oxfordian-Kimmeridgian boundary zone, Neuffen, South Germany.
F: Cretaceous-Tertiary boundary, Scaglia Rossa, Contessa valley, Gubbio, Italy. A bleached zone 30cm thick developed below the boundary clay. Note frequency of stylolitic bedding planes.
G,H: Marl-limestone alternation at the transition between the Scaglia Rossa/Scaglia Cinerea showing frequent stylolitic bedding planes. Eocene and Oligocene, Contessa valley, Gubbio, Italy.

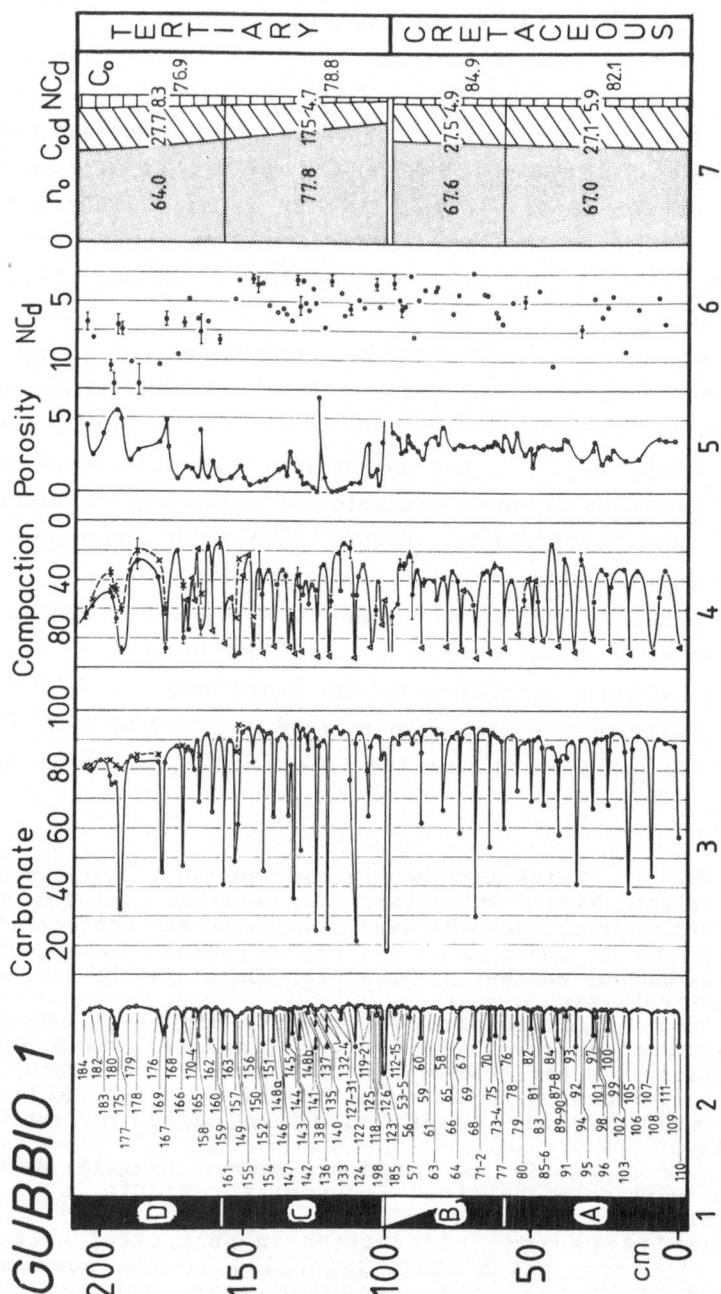

Fig. 37 Gubbio 1 marl-limestone alternation spanning the Cretaceous-Tertiary boundary, Scaglia Rossa, Umbria, Italy. Columns: 1) reddish rock (black), light grey rock (white), 2) weathering profile and sample numbers (Maastrichtian was sampled in the Bottaccione gorge, Paleocene was sampled in the Contessa valley, 3) carbonate content of selectively cemented burrows (x) and rock matrix (•), 4) compaction of

cementation at the Cretaceous-Tertiary boundary, as ARTHUR (1979)
infers from the 30 to 50cm thick bleached zone below the boundary
clay. Bleaching of the red iron pigment could be explained by oxygen
depletion during deposition of the boundary clay.

Calculations and results: The average normalized noncarbonate
fraction (eq. 2) decreases upsection from 6 to 5% towards the
Cretaceous-Tertiary boundary and then increases to more than 8% in the
Tertiary section (Fig. 37). Contrary to the behavior of porosity
which is the result of differential cementation, the standardized
noncarbonate fraction remains unchanged directly at the boundary.
Unfortunately, compaction measurements could not be computed for the
boundary clay because individual burrows could not be detected. Since
compaction was evaluated predominantly for the limestone layers, it is
difficult to determine the primary compositional differences between
the existing marl and limestone layers. Only within subsection D
could the absolute clay content be calculated for lithologies of
varying carbonate content (Table 9). Because the mean absolute clay
content in subsection D is nearly constant, it was presumed that the
absolute clay content in subsections A through C fluctuates only
slightly. Cretaceous (subsections A,B) and Tertiary rocks
(subsections C,D) were then decompacted and balanced (Figs. 38d,e;
39a,b).

Table 9: Mean standardized noncarbonate fraction
(Gubbio 1 section, subsection D).

Carbonate content of the rock (%)	20-80	80-100
Mean standardized noncarbonate fraction (NC_d, vol %)	8.20	8.30
Number of measurements	8	9

◁ selectively cemented burrows (x) and rock matrix (●),
interpolated values of compaction using eq. 4 and mean
absolute clay contents (Δ), 5) porosity, 6) noncarbonate
fraction, expressed as a percentage of the primary or
decompacted sediment volume (NC_d), 7) mean original sediment
composition with the primary porosity (n_o), and absolute
values of the original carbonate content (C_{od}) and
noncarbonate content (NC_d), 8) primary relative carbonate
content (C_o).

GUBBIO 1

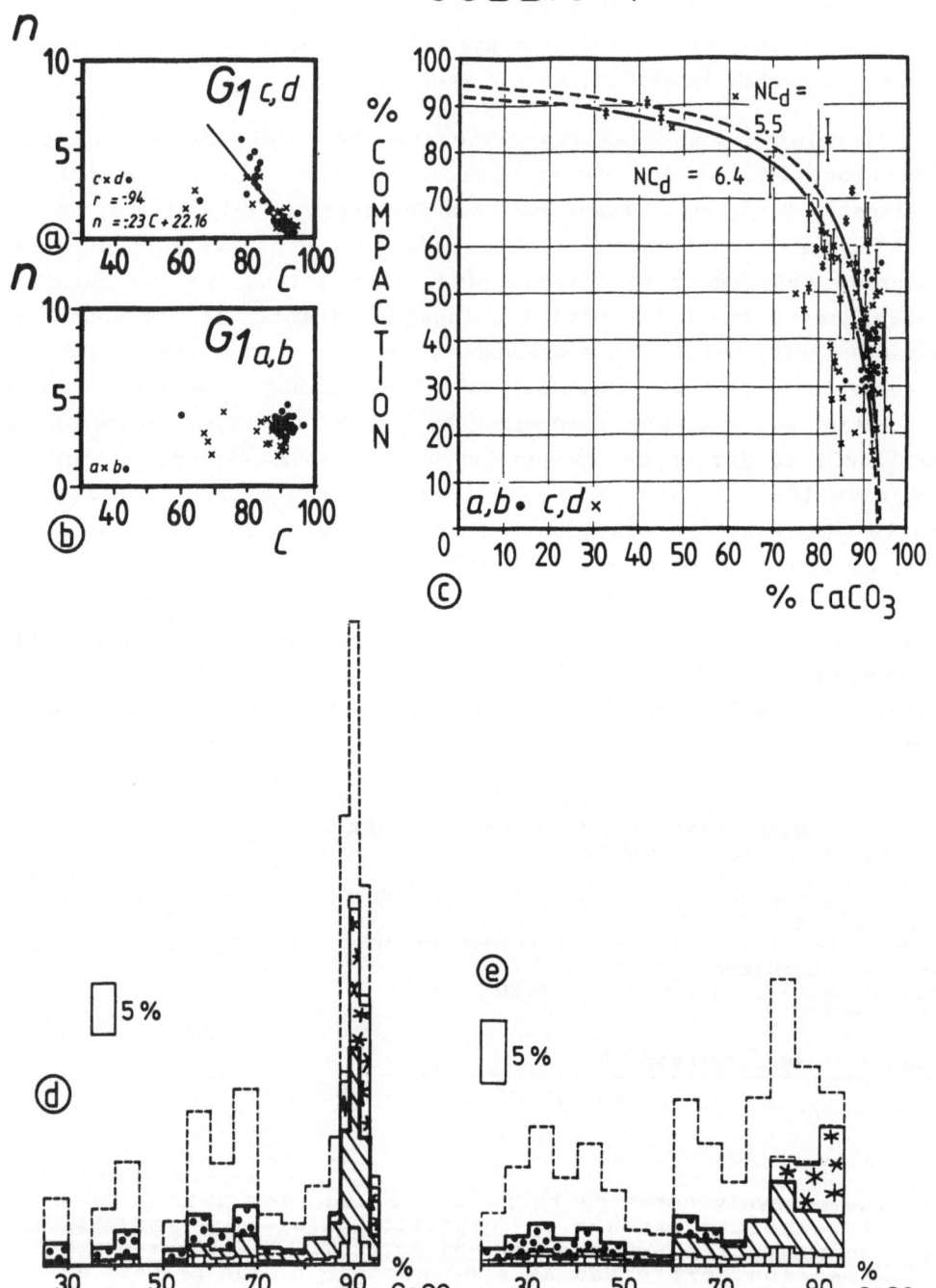

The results of these calculations indicate that mean decompaction porosities are 67% in the Cretaceous and 71% in the Tertiary subsections with mean primary carbonate contents of 83 and 77%, respectively. In 57 to 60% of the original sediment column, carbonate was diagenetically released and then reprecipitated in the cementation zones (Fig. 39a,b). Therefore, one must assume that the deposition of the primary sediment, which is represented by the present, stylolitic seams, lasted somewhat longer than the amount of time needed for the deposition of the limestone layer sediment. ARTHUR (1979) has already reached similar conclusions. Given that porosity varied by 5% in the original sediment, mean primary carbonate differences between the layers which later become dissolution and cementation zones (see section 2.3.7) were 2.5% (Maastrichtian) and 3.9% (Paleocene). Present mean carbonate variations are 25.8% (Maastrichtian) and 31.1% (Paleocene); thus, diagenetic enhancement is by a factor of 10.3 and 8.0 for the Maastrichtian and Paleocene, respectively. Consequently, the limestone layers contain relatively large quantities of cement (an average of 46 to 48% of the total carbonate fraction). The cement is derived from the stylolitic marl seams where 64% of the primary carbonate is dissolved.

Compaction at the onset of the cementation processes is 33.5% in the Maastrichtian and 28.2% in the Paleocene section (Fig. 39c,d). This corresponds to a mean overburden at the onset of cementation of 200m (Maastrichtian) and 120m (Paleocene), ranging from 100 to 450m (Maastrichtian) and from 65 to 220m (Paleocene). Mechanical compaction lasted an average of 22 and 16 million years for the Cretaceous and Tertiary subsections, respectively. Calculations were performed using the following decompacted sedimentation rates (according to compacted sedimentation rates from ARTHUR & FISCHER, 1977) in m/1000 years: Paleocene, 0.01; Eocene, 0.02; and Oligocene, 0.03.

Discussion: The following summary results from independent carbonate mass balance calculations for the four subsections (A,B,C

Fig. 38 Data from the Gubbio 1 section with subsections A through C.
a,b: Relationship between porosity (n) and carbonate content (C); sections A,B (b), section C,D (a).
c: Measurements and theoretical curve calculated from the compaction law using mean absolute clay contents of 5.5% (sections A,B) and 6.4% (sections C,D) and the formula for porosity used in Fig. 38a,b.
d,e: Carbonate mass balance, sections A,B (d), sections C,D (e).

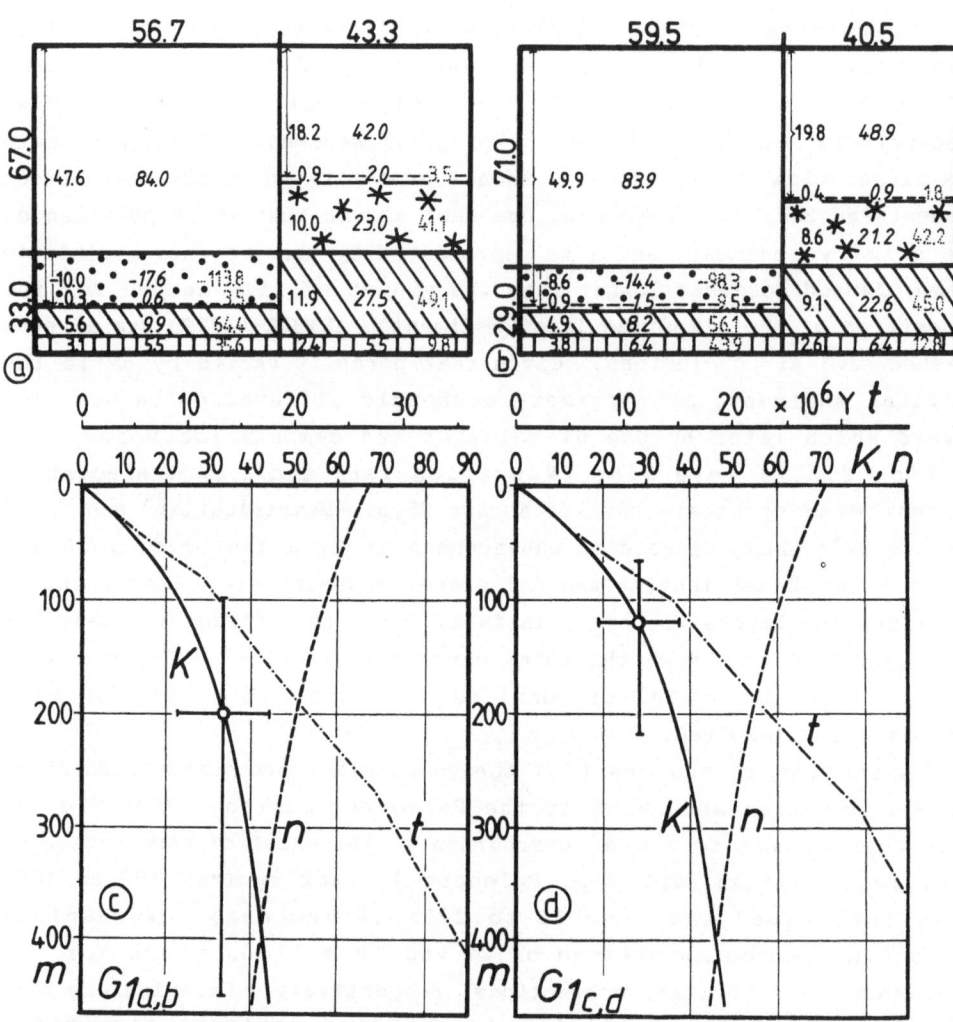

Fig. 39 Data from Gubbio 1 section with subsections A through D.
a,b: Carbonate mass balance showing box models; sections A,B (a), sections C,D (b).
c,d: Porosity (n), compaction (K), and time (t) in the sediment column, based on DSDP data (HAMILTON, 1976). Value shown represents compaction at the onset of lithification with a $\pm\sigma$ zone of scatter; sections A,B (c), sections C,D (d).

and D; Fig. 37). First, during the uppermost Maastrichtian, calcareous ooze, with an average carbonate content of 82 to 85% and a mean decompaction porosity of 67 to 68%, was deposited. After the deposition of the boundary clay (that is, after the iridium event), mean porosity increased to 78% for an interval ranging from 10,000 to 100,000 years and then decreased to 64%, while the mean carbonate content dropped suddenly by 6%. The high porosity in the lowermost

Paleocene presumably results from a greater sedimentation rate after the iridium event. Moreover, Paleocene sediments must have had a greater proportion of unstable carbonate phases (e.g., aragonite, Mg-calcite), because lithification of these sediments began 6 million years earlier and their average overburden was 80m less compared to the Maastrichtian. Therefore, the Paleocene limestones have lost most of their original porosity during cementation.

3.5.2 Stylolitic Alternation, Oligocene (G2)

Location: Transition zone between the Scaglia Rossa and Scaglia Cinerea in the Contessa valley.

Section: Major carbonate oscillations weather to limestones which are 15cm thick and are separated by narrow marl beds (Fig. 40). Smaller carbonate cycles are superimposed on this major rhythm due to pressure dissolution planes (Fig. 36G,H). The carbonate content is between 90 and 95% in the limestones, and it ranges from 50 to 80% in the marl beds and stylolitic bedding planes. Compaction (having been both directly and indirectly evaluated using Planolites and Teichichnus) varies from 30% in the limestone layers to 80% in the marl beds. Unfortunately, compaction usually could not be ascertained in the stylolitic marl joints.

The mean concentration of Mn in the carbonate fraction (660ppm) is 4 to 5 times greater than and the concentration of Fe (390ppm) and Mg (1800ppm) are 2 to 6 times less than in the hemipelagic and epicontinental alternations of Angles 2 (section 3.3.2) and of Neuffen 2 (section 3.4.2). Therefore, these minor elements trapped in the carbonate fraction (Fig. 40) reflect the deep-water environment of the Scaglia limestones. For the most part, the concentration of the minor elements is inversely proportional to the carbonate content and directly proportional to the degree of compaction. Mg, Fe, Sr, and Mn were enriched in the marl beds relative to the amounts contained in the limestone layers by maximum factors between 1.2 and 2.

Calculations and results: The normalized noncarbonate fraction (or the absolute clay content, eq. 2) range from 4 to 8% (Figs. 40, 41b). It exhibits no significant changes when the carbonate content is more than 80% (Table 10). Nevertheless, the absolute clay content decreases slightly with lower carbonate content; this might indicate that the original sediment which is now represented by the marl beds

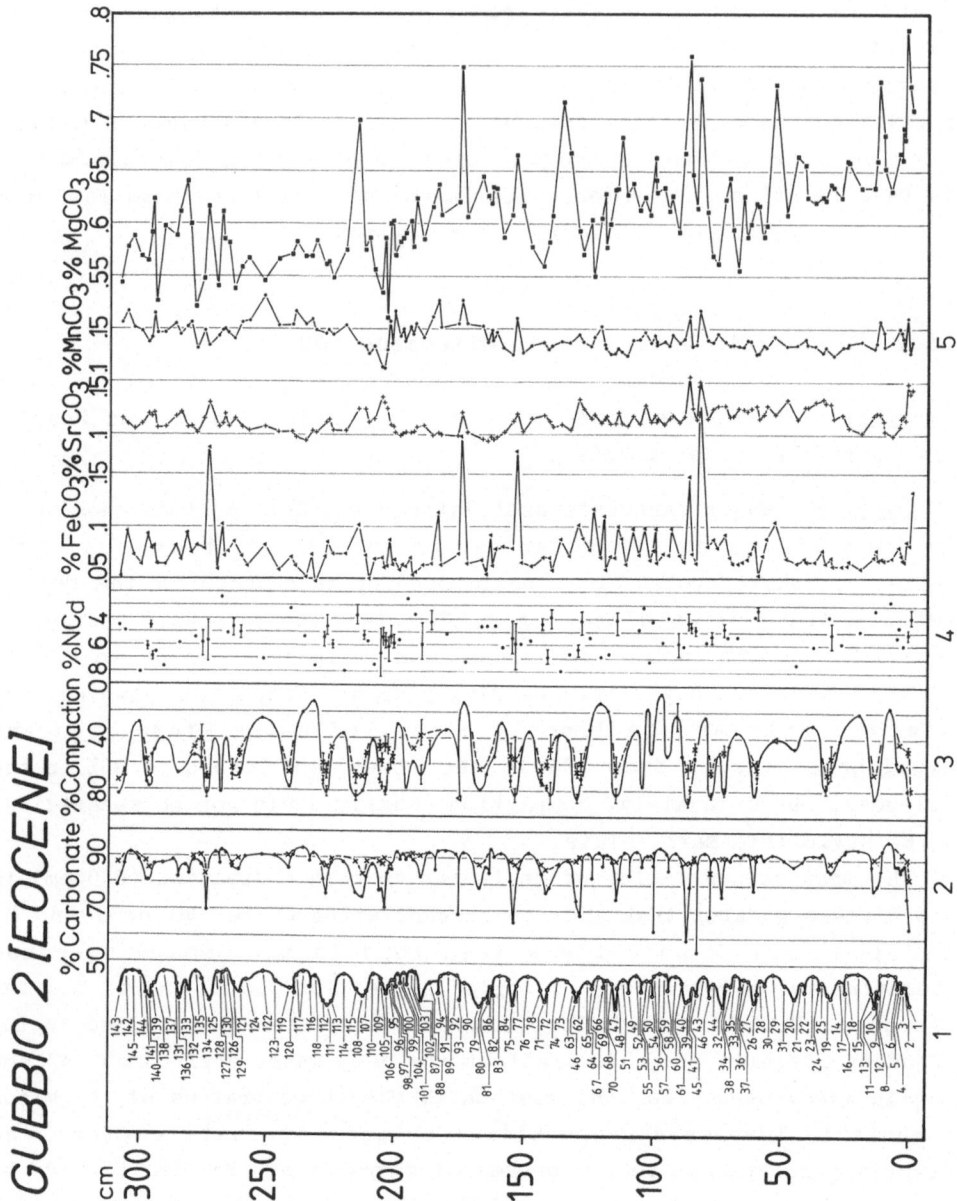

Fig. 40 Gubbio 2 marl-limestone alternation, Eocene, Contessa valley, Italy. Columns: 1) weathering profile and sample numbers, 2) carbonate content in selectively cemented burrows (x) and in the rock matrix (•), 3) compaction in selectively cemented burrows (x) and in the rock matrix (•), 4) noncarbonate fraction, expressed as a percentage of the original sediment volume (NC_d), 5) trace and minor elements, expressed as a percentage of the total carbonate fraction.

had a slightly higher primary porosity than that of the present limestone layers (see section 2.3.4). Decompaction and carbonate mass balance calculations (Fig. 41c,d) yield a mean primary porosity of 66.4% and an original carbonate content of 85%.

Table 10: Mean standardized noncarbonate fraction (Gubbio 2 section).

Carbonate content of the rock (%)	75	75-80	80-85	85-90	90-95
Mean standardized noncarbonate fraction (NC_d, vol %)	4.73	4.49	5.21	5.26	5.21
Standard deviation	–	0.93	0.55	1.57	1.06
Number of measurements	9	13	10	32	24

An average of 41% of the primary carbonate content in the marl layers was dissolved to produce carbonate cement which comprises 34% of the carbonate in the limestone layers. If a porosity variation of 5% in the primary sediment is assumed, original mean carbonate differences are 2.3% (83.6% for dissolution zones and 85.9% for cementation zones). Since present mean carbonate differences are 12.8% (76.7% for dissolution zones and 89.5% for cementation zones), carbonate fluctuations were enhanced to 5.5 times their original value.

Compaction at the onset of lithification had an average amount of 27.3% (Fig. 41d). According to eq. 8 and the methods presented in section 2.4, porosity at the beginning of cementation was 54% and mean overburden was 138m (the standard deviation of K_1 values indicate that overburden was between 70 and 270m). Mechanical compaction lasted about 6.3 million years (decompacted sedimentation rates were calculated using compacted sedimentation rates from ARTHUR & FISCHER, 1977): Oligocene, 0.03; Miocene, 0.21; in m/1000 years).

3.5.3 Marly to Silty Alternation, Oligocene (G3)

Location: Road cut in the Contessa valley near entrance of the upper quarry.

Section: A random spot sampling consisting of 32 samples was made in an interval of 30 meters; the section is a marly to silty alternation of the Scaglia Cinerea. Compaction was mostly evaluated

GUBBIO 2

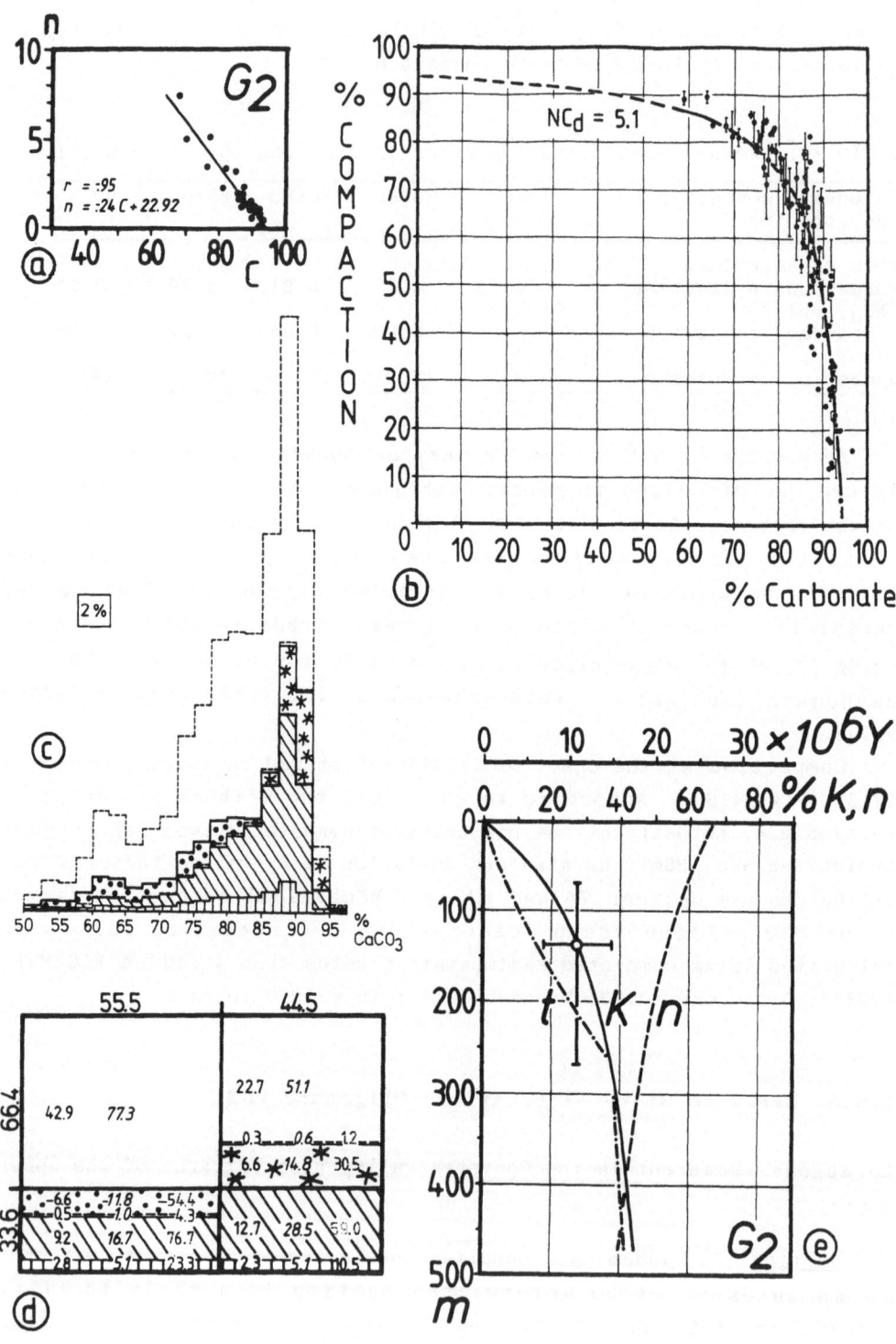

from selectively early cemented burrows using the indirect method introduced in section 2.1.2. Assuming that the samples in the random spot sampling truly represent the bedding variations, the primary sediment had a mean porosity of 66%, an average carbonate content of 75%, and an absolute clay content of 8.4%.

Discussion: Although Upper Maastrichtian and Paleocene limestones from the Gubbio section have various primary carbonate contents of 75 to 85%, the decompaction porosities are independent of the carbonate content and range from approximately 64 to 68% (Fig. 42). Only the lowermost Paleocene seems to be an exception; where primary porosities increased to 78%. Therefore, this provides documentation of the high sedimentation rates directly following the Cretaceous-Tertiary boundary event (see section 3.5.2).

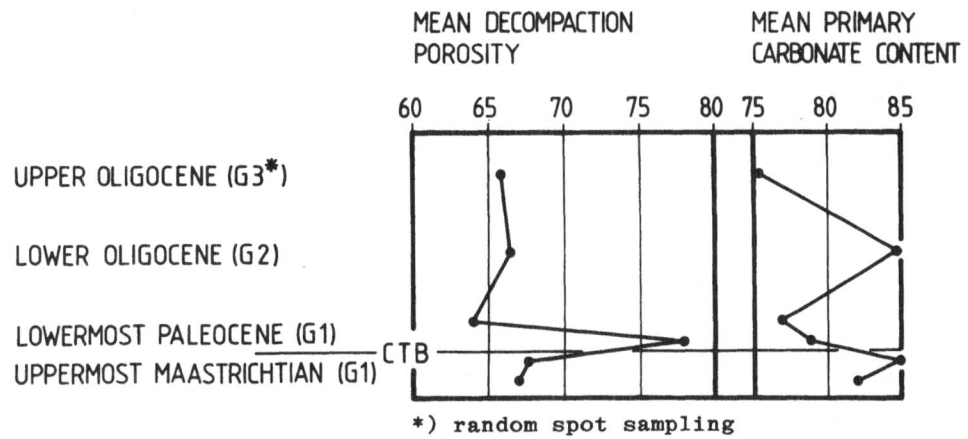

*) random spot sampling

Fig. 42 Mean amounts of decompaction porosity and primary carbonate at several sites in the Gubbio section (Italy). Note the relatively high value of the decompaction porosity directly above the Cretaceous-Tertiary boundary (CTB).

◁**Fig. 41** Data from the Gubbio 2 section.
a: Relationship between porosity (n) and carbonate content (C).
b: Measurements and theoretical curve calculated from the compaction law using mean absolute clay content of 5.1% and the formula for porosity used in Fig. 41a.
c,d: Carbonate mass balance showing histogram and box model.
e: Porosity (n), compaction (K), and time (t) in the sediment column, based on DSDP data (HAMILTON, 1976). Value shown represents compaction at the onset of lithification with a $\pm\sigma$ zone of scatter.

Table 11: Compilation of data in the marl-limestone alternations studied.

Section	PE	R	A1	A2	A3	L	N1	N2	G1A,B	G1C,D	G2
Type of alternation	O	I	I	II	III	II	I	III	II-III	II-III	II
Maximum sedimentary overburden (m)	500	1900	1250	950	900	1500	>400	>400	>2500	>2500	>2500
Mean overburden at the onset of lithification (m)	>500	276	460	172	135	169	130	80	200	120	138
Mean compaction at the onset of lithification (K_1,%)	36.2	32.9	54.1	36.0	26.4	27.9	27.4	19.3	33.5	28.2	27.3
Mean porosity at the onset of lithification (n_1,%)	--	38.2	50.8	58.1	52.9	47.4	55.5	57.0	50.4	59.6	53.8
Mean decompaction porosity (n_0,%)	64	58.5	77.4	73.2	65.3	62.1	67.7	65.3	67.0	71.0	66.4
Mean standardized non-carboante fraction (NC_d, vol %)	13.2	10.7	6.0 7.4	6.4	3.4 8.5	7.3 7.2	6.3 7.0	3.2 4.0	5.5	6.4	5.1
Mean primary carbonate content in dissolution and cementation zones (C_0,%) *)	--	72.6 75.4	63.2 76.1	73.7 78.2	73.6 90.9	79.4 81.9	76.5 81.9	87.6 91.4	82.0 84.5	75.8 79.7	83.6 85.9
difference	--	2.8	12.9	4.5	17.3	2.5	5.4	3.8	2.5	3.9	2.3
Mean existing carbonate content in dissolution and cementation zones (C,%)	--	62.9 80.6	60.5 80.4	63.4 83.9	47.9 94.1	74.0 86.8	71.1 86.8	77.2 94.0	64.4 90.2	56.1 87.2	76.7 89.5
difference	--	17.7	19.9	20.5	46.2	12.8	15.7	16.8	25.8	31.1	12.8

Section	PE	R	A1	A2	A3	L	N1	N2	G1A,B	G1C,D	G2
Factor of diagenetic enhancement (F) *)	--	6.3	1.5	4.6	2.7	5.1	2.9	4.4	10.3	8.0	5.5
Mean relative cement content (Z_c,%) of the cementation zones	--	30.8	32.6	38.9	41.9	35.9	36.6	38.8	45.6	48.4	34.1
Volume of the primary sediment which underwent cementation (V_z,%)	--	47.9	31.9	41.5	44.9	36.5	34.8	46.4	43.3	40.5	44.5
Carbonate content of the neutral value (C_n,%)	--	73.8	70	76	90	81	79	89	83	78	85
Carbonate content at the weathering boundary (C_w,%)	60	72-77	72	74-82	80-90	85	85	90	80	85-90	82-88

PE = Porto Empedocle; R = Rheine; A1-3 - Angles 1 to 3; L = Logis du Pin; N1-2 = Neuffen 1 and 2; G1A,B and G1C,D = Gubbio 1, subsections A to D; G2 = Gubbio 2.
*)assuming a variation in the primary porosity of 5%.

3.6 Results and Bedding Types

1. As already discussed in section 2.3.1, diagenetic bedding must occur within a closed or nearly closed carbonate system, because the decompaction calculation gives realistic mean primary porosities ranging from 59 to 78% (see Fig. 10). However, until the onset of diagenetic carbonate redistribution sediments were mechanically compacted under an overburden of 50 to 450m. Mechanical compaction was usually greater when the primary sediment had a lower carbonate content (Table 11, see Fig. 60a).

2. Primary sediments were calcareous oozes with original mean carbonate contents between 60 and 90%. After sediment mixing due to bioturbation <u>primary mean carbonate differences between dissolution and cementation zones</u> were between 2.3 and 17.3% with an average value of 5.8%. According to section 2.3.7 (Fig. 14),

these values have to be multiplied by a factor ranging from 1.5 to 2 when the average fluctuation between the maxima and minima on the carbonate curve is calculated. The primary carbonate variations after bioturbation were diagenetically enhanced by factors of 1.5 to 10.3 (average enhancement was by a factor of 5.1). Diagenetic bedding formed parallel to the original stratification.

3. In the primary sediment column, beds which underwent dissolution and those which underwent cementation did not have the same thickness. Only an average of 41% of the primary sediment volume became cemented. Nevertheless, in carbonate-rich alternations as they appear today, limestone layers comprise most of the rock column, whereas the thickness of the marl beds is reduced by 60 to 80% of their original thickness as a result of mechanical and chemical compaction. Therefore, diagenesis affects not only carbonate variations but also bedding rhythmicity.

4. When plotted, carbonate curves of lithified marl-limestone alternations are usually not sinusoidal as is the case for prediagenetic alternations. Depending on the total carbonate content of the primary sediment and the different changes in volume due to carbonate redistribution processes, carbonate curves become convex to angular in the limestone layers and narrow in the marl beds. Moreover, the carbonate content in the middle of the limestone layers always exhibits smaller variations than does the minimum carbonate content in the marl layers.

5. In the lithified alternations studied, no close relationship exists between porosity and overburden. Nevertheless, the evidence clearly demonstrates that rock porosity depends on carbonate content. In lithified alternations, porosity increases with decreasing carbonate content (see Fig. 62). Whereas in unlithified alternations the behavior is the direct opposite.

6. As already introduced in Fig. 14, dissolution and cementation zones do not correspond exactly to marl and limestone beds in the weathered section. Only at low absolute clay values (about 3%) or high mean carbonate contents do the thickness of dissolution and cementation zones equal the thickness of the weathered marl and limestone layers (Fig. 43). However, in clay-rich alternations, the cemented portions of the alternations are thicker than the weathered limestone ledges in the outcrop.

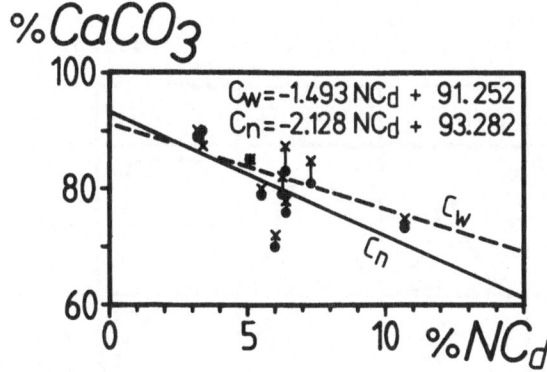

Fig. 43 The carbonate content at the weathering boundary (C_W, x, ---) and the neutral carbonate value (C_n, •, ——) in relation to the standardized noncarbonate fraction (NC_d).

According to the different forms of the carbonate curves, four basic types of marl-limestone alternations can be distinguished. One of these is unlithified (Type O), while the remaining three are lithified (Types I to III; Table 12):

Type O Unlithified Alternations

Unlithified alternations are still in the mechanical compaction phase. They show only weak to moderate sinusoidal or complex carbonate oscillations due to stochastic depositional processes, depositional events, and bedding cycles. They are usually bioturbated, and they commonly have mean carbonate differences below 15% (calculated between the maxima and minima on the carbonate curve; see Fig. 14). Porosity decreases with decreasing carbonate content. Unlike lithified sediments, which follow the theoretical curves of the carbonate compaction law for low porosities (see Fig. 7), unlithified alternations display only moderate differences in carbonate content and compaction. Examples: the calculated original sediments (section 3.2 to 3.5) and the Porto Empedocle section describing uncemented foraminiferal ooze (section 3.1).

Table 12: Carbonate curves of marl-limestone alternations. Type 0 is prediagenetic; Types I, II, and III are lithified.

	TYPE 0	TYPE I	TYPE II	TYPE III	
		sinusoidal	convex	angular	SHAPE OF THE CARBONATE CURVES
					MEASURED CARBONATE COMPACTION DATA
		35 - 55	25 - 35	15 - 25	COMPACTION AT THE ONSET OF LITHIFICATION IN %
		6 - 10	4 - 6	3 - 4	STANDARDIZED NON-CARBONATE FRACTION IN %
		30 - 35	35 - 40	40 - 48	% CEMENT (CEMENTATION ZONE) RELATIVE TO THE TOTAL CARB. CONTENT
		1 - 6	4 - 10	3 - 5	FACTOR OF ENHANCE-MENT OF PRIMARY CARB. DIFFERENCES
		60 - 80	75 - 90	85 - 91	MEAN PRIMARY $CaCO_3$ CONTENT IN %

Type I Lithified Alternations with Carbonate Curves

Characterized by Sinusoidal Maxima

These are alternations relatively rich in marl with absolute clay contents between 6 and 10% and original carbonate contents between 60 and 80%. Mechanical compaction prior to the onset of lithification was high (35 to 55%), and the pore space to be cemented was considerably reduced. Therefore, both the cement content in the cementation zones (30 to 35% of the total carbonate content) and the degree of diagenetic enhancement are relatively low. For that reason, measured data scatter around only a small part of the theoretical curve representing the relationship between carbonate content and compaction (Table 12). Carbonate content in the limestone layers plots sinusoidally with maximum carbonate contents of 80 to 85%. The neutral carbonate value (that is, the carbonate content between dissolution and cementation zones) is still within the weathered marl layers. Examples: studied sections from Rheine (section 3.2), Angles 1 (section 3.3.1), and Neuffen 1 (section 3.4.1).

Type II Lithified Alternations with Carbonate Curves
Characterized by Convex Maxima

This type of marl-limestone rhythm is characterized by broad, convex carbonate peaks with maximum carbonate values of 85 to 95%, medium absolute clay contents of 4 to 6%, medium amounts of cement (35 to 40% of the total carbonate in the limestone layers), but high levels of diagenetic enhancement (from 4 to 10 times the original $CaCO_3$ variations). Commonly, the neutral carbonate value is slightly lower than the carbonate content at the weathering boundary. Examples: studied sections of Angles 2 (section 3.3.2), Logis du Pin (section 3.3.4), and partly Gubbio 1 and 2 (sections 3.5.1 and 3.5.2).

Type III Lithified Alternations with Carbonate Curves
Characterized by Angular Maxima

The limestone layers of this highly conspicuous type of alternation have broad, angular carbonate curves; maximum carbonate values reach more than 95%. Carbonate content diminishes abruptly just prior to the contact with the marl beds. Usually, the marl beds are only a few centimeters thick; thus, limestone layers predominate in the weathered section, resulting in a brick appearance where the limestones are the bricks and the marl represent the grout between them. In the original sediment, carbonate content increased to more than 85%. Due to this, the reduction of the original pore space, caused by mechanical compaction, was low; and, therefore, the cement content in the limestone layers is high (40 to 48% of the total carbonate content in the limestone layers). Nevertheless, original carbonate variations were enhanced by a factor from only 3 to 5. Measured data correspond relatively well to the right branch of the theoretical curve expressing the relationship between carbonate content and compaction (see Fig. 7 and Table 12). The carbonate neutral value (between the dissolution and cementation zones) is usually higher than the carbonate content of the weathering boundary; thus, cementation zones are thinner than the weathered limestone layers. Therefore, limestone layers always have a flasery-stylolitic outer edge. Pressure shadow structures around massive shells (see Fig. 58) are frequently found in the marl joints. Examples: section Angles 3 (section 3.3.3) and Neuffen 2 (section 3.4.2).

4 DIAGENETIC LEDGE FORMATION AND BEDDING RHYTHMICITY

In the previous section, the production of diagenetic bedding was quantified through the use of various examples. In the following sections, the origin of different carbonate curves and diagenetic rhythmicity as expressed in calcareous bedding alternations is studied.

4.1 Carbonate Profiles in Limestone Layers

As can be seen in Table 12, the development of a particular type of carbonate curve is dependent on the absolute clay content and on the degree of mechanical compaction. The carbonate curve displays a sinusoidal shape when the absolute clay content and the degree of mechanical compaction are high. Conversely, the $CaCO_3$ curve shows angular maxima (in the limestone layers) and pointed, sharp minima (in the marl beds) when the absolute clay content and the degree of mechanical compaction are low.

This phenomenon can be explained in a simple manner using compaction models for limestone layers. Fig. 44 presents a hypothetical sediment column which is divided into subsections of equal thickness. In the model one assumes that compaction, which is lowest in the middle of what later becomes the cementation zone of the limestone layer, progresses steadily outwards during diagenesis. Therefore, the thicknesses of the subsections become smaller with increasing compaction. This model produces the typical compaction pattern found in this study. Compaction increases markedly from the middle of the limestone layer (the center of cementation) towards the marl beds. The distribution of compaction explains why layers with different absolute clay content have varying shapes of their carbonate curves, since compaction can be transformed into carbonate content by using the compaction law and various mean absolute clay contents.

Fig. 44a,b present simple constructions of the above-mentioned carbonate curves for limestone layers. Mechanical compaction is neglected. When the absolute clay content is high ($NC_d=10\%$, Fig.

CARBONATE DISTRIBUTION IN LIMESTONE LAYERS

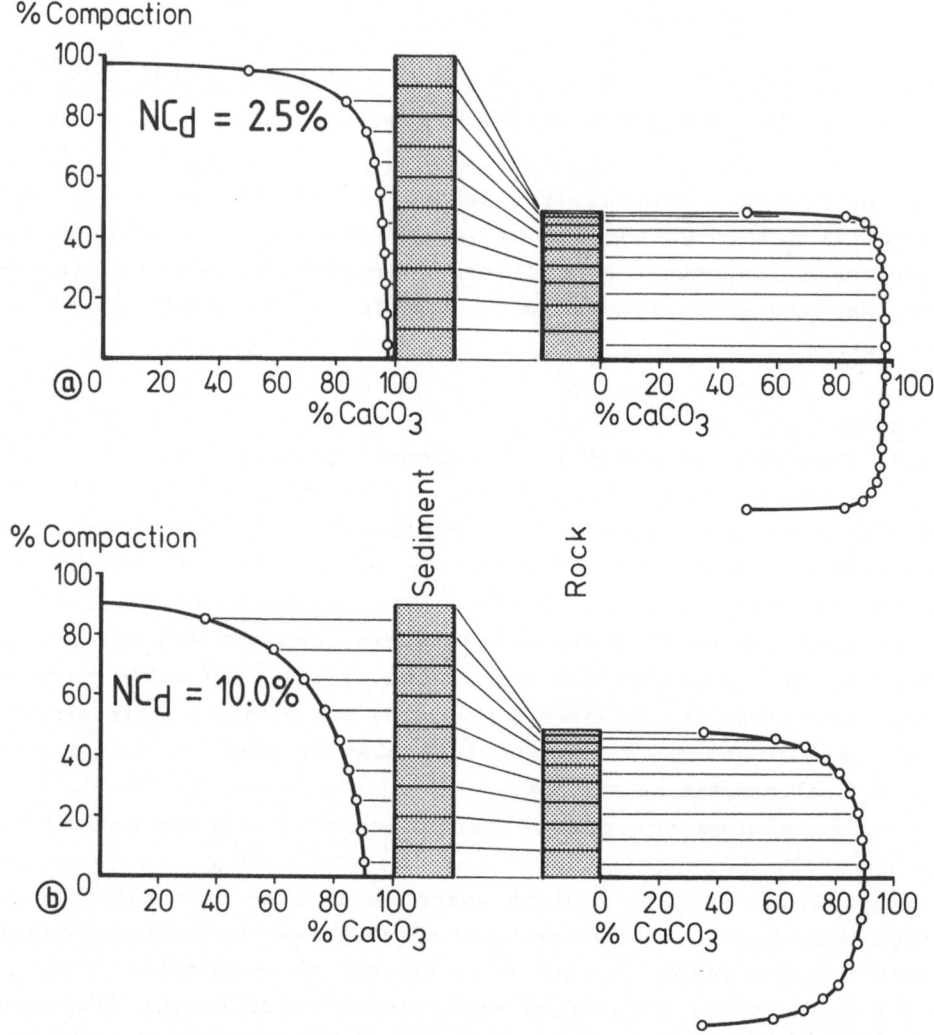

Fig. 44 Schematic models which explain the different types of carbonate curves in limestone layers. The models use the carbonate compaction law with absolute clay contents of $NC_d=2.5$ and 10%. The porosity in the existing limestone layers is assumed to be negligible.
a,b: The mechanical compaction prior to the onset of cementation in the middle of the limestone layer is assumed to be zero.
c,d: Mechanical compaction in the middle of the limestone layer is supposed to be MK=35%.

44a), the compaction law (eq. 4) simply produces a curved relationship between carbonate content and compaction. According to this relationship, one obtains sinusoidal carbonate curves for the

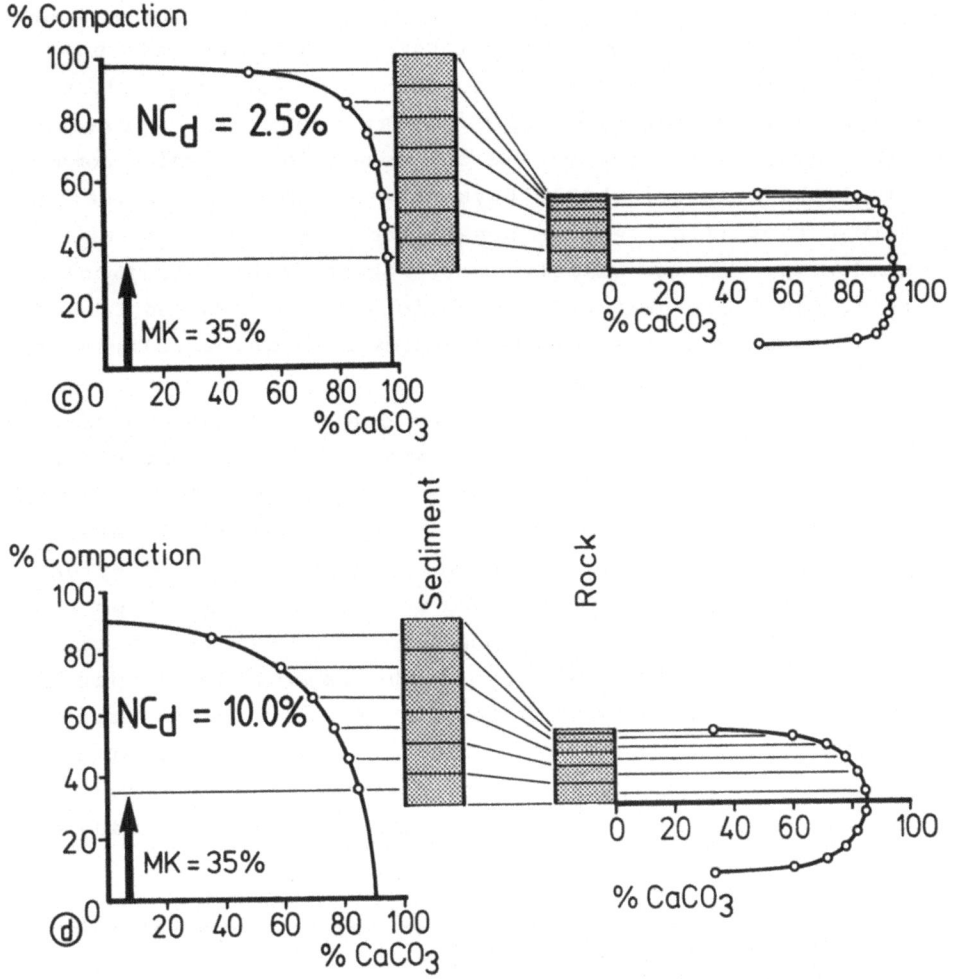

limestone layers in marl-rich alternations. On the other hand, when the absolute clay content is low (NC$_d$=2.5%, Fig. 44b), the dependence of carbonate content on the increasing degree of compaction is at first negligible. Only at high values of compaction does this dependency exist. Therefore, at low absolute clay contents, the carbonate compaction law displays a curve with an angular shape (see Fig. 7). For that reason, limestone layers in carbonate-rich alternations have angular shapes of their carbonate curves.

The previously-described models are further modified by assuming that mechanical compaction in the middle of the limestone layer is 35% until the onset of cementation (Fig. 44c,d). It is evident from this model that mechanical compaction causes the shapes of the carbonate curves in limestone layers to become narrower and more curved. The greater the amount of mechanical compaction, the more the carbonate distribution in the limestone layer is determined by the left portion of the theoretical carbonate-compaction curve (see Fig. 7), where compaction greatly affects the carbonate content. Therefore, limestone layers with high mechanical compaction (or which commonly begin to lithify later) should have narrower and more sinusoidal carbonate curves with lower carbonate contents than layers with low mechanical compaction and usually early beginning of lithification. This relationship is particularly obvious in the Neuffen 1 section (see Fig. 32), in which the carbonate maxima plot more angularly when the carbonate content increases.

In the following pages, the previous models become further quantified. First, it shall be tested whether or not compaction in the limestone layers increased evenly at every step of lithification. In order to detect this, compaction in several limestone layers form the Angles and Neuffen sections was calculated (eq. 4) by using detailed carbonate curves and mean absolute clay contents from those sections. Then the thickness of each limestone layer was decompacted according to eq. 7 (Fig. 45a). If the amount of compaction continuously increased throughout the primary sediment, the compaction curve calculated for the decompacted sediment column would be straight. However, limestone layers do not conform exactly to this curve. Compaction (Fig. 45a) decreases only slightly in the middle of

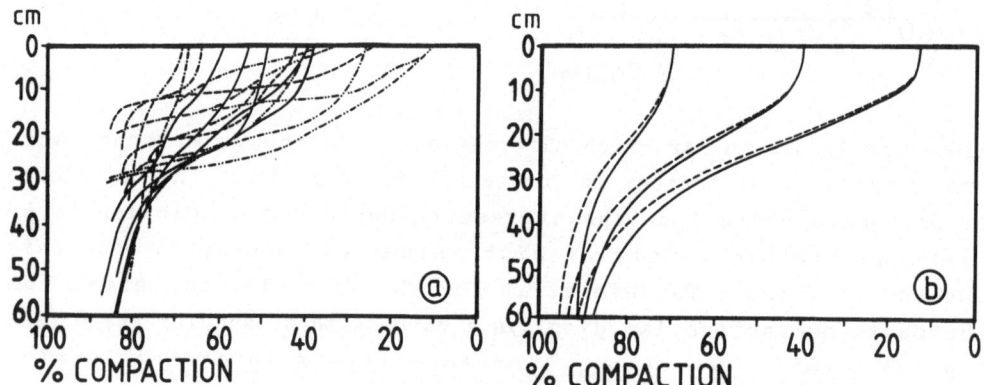

Fig. 45 Increasing degree of compaction from the middle of some limestone layers towards their marl beds versus the thickness of the decompacted sediment column.
a: Limestone layers of Angles 1 (−·−), Angles 2 (———), Angles 3 (−−−), and Neuffen 2 (−··−).
b: Simulated compaction curves shown for absolute clay contents of 2.5% (−−−) and 7.5% (———), as expressed in eq. 11, with values for the compaction at the onset of lithification, K_1=12.5%, 40%, and 70%.

the limestone layers; this may be explained by the intensity of cementation processes during the beginning of lithification.

The different compaction curves in Fig. 45a were simulated using the empirical equation

$$h^* = \left(\frac{K-K_1}{100-K-NC_d}\right)^{1/3} \times 25 \quad , \quad (11)$$

which fits the curves well (Fig. 45b); where h^* is the decompacted, original sediment thickness, K and K_1 are the amount of compaction and the amount of compaction at the onset of lithification, respectively, and NC_d is the clay content normalized to the decompacted sediment volume.

The different types of carbonate curves can be calculated as follows: Primary sediment intervals (Δh^*) must be continuously compacted (eq. 11). Thereby, the process begins in the middle of the later limestone layer. The intervals have to be compressed reflecting the specific amount of their compaction and then added to give the compacted (half) limestone layer thickness. For every value calculated for compaction the carbonate content is then determined by using eq. 5 with a constant mean absolute clay content.

The resulting plots describing carbonate curves in limestone layers for the Angles and Neuffen sections are presented in Fig. 46, where the parameters are the absolute clay content (NC_d) and the degree of compaction (K). Fig. 46 confirms the previously-stated opinion that limestone layers have angular carbonate curves when the absolute clay content is low and have sinusoidal carbonate curves when the absolute clay content is high. Moreover, the shapes of the carbonate curves and their maximum carbonate content depend on the degree of mechanical compaction, since the carbonate peaks become smaller, narrower, and less pronounced as mechanical compaction increases.

In the alternations studied, the influence of the absolute clay content on the shapes of the carbonate curves is enhanced in such a manner that the degree of mechanical compaction increases with increasing absolute clay content in the primary sediment (see Table 12 and Fig. 60). In simpler terms, sediments which contain a high primary carbonate content and a low absolute clay content usually undergo slight to moderate degrees of mechanical compaction and

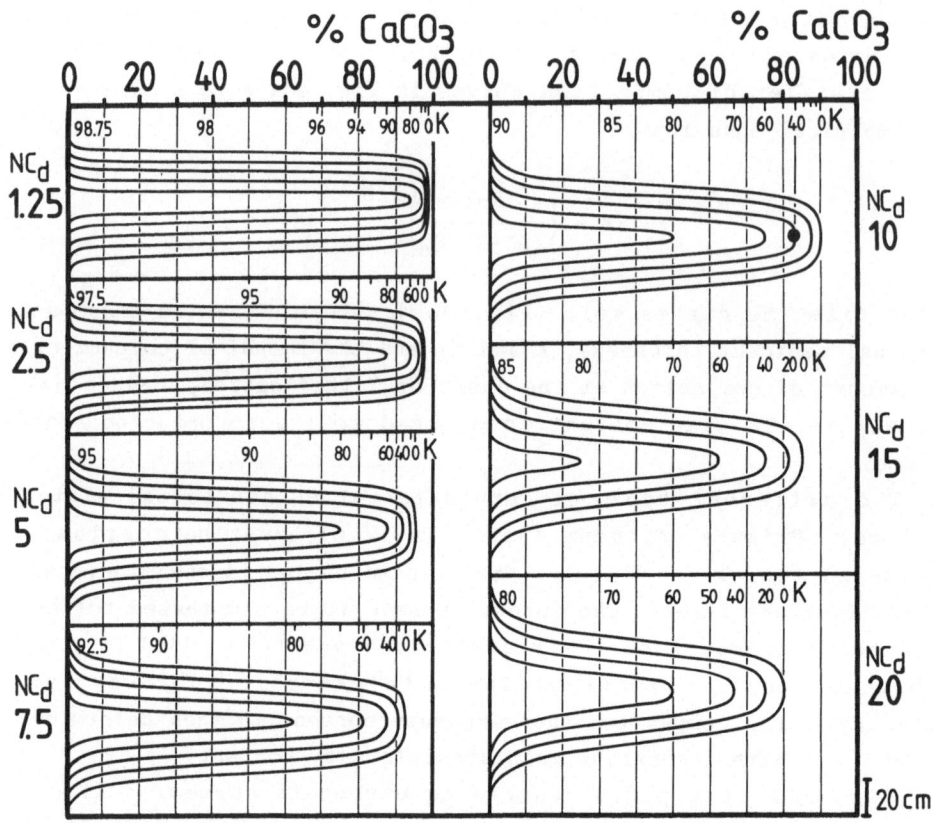

Fig. 46 Simulated carbonate curves for limestone layers in some Jurassic and Cretaceous sections from southeastern France and southern Germany. The carbonate curves depend on the normalized noncarbonate fraction (NC_d=1.25 to 20%) and on the amount of compaction (K, small numbers at the nonlinear scales). It is assumed that porosity in the limestone layers is negligible.
Example: Carbonate curve for a limestone layer with a standardized noncarbonate fraction of 10%, a maximum carbonate content of 83%, and a mechanical compaction at the onset of lithification of 40%.

commonly begin to lithify early; thus, diagenesis forms carbonate peaks with angular shapes. However, in initially, clay-rich sediments, the onset of lithification is comparatively later and occurs when mechanical compaction is greater. Therefore, in clay-rich sequences, carbonate diagenesis only generates peaks with sinusoidal, narrow shapes.

4.2 Thickness of the Limestone Layers

Since the absolute clay content is approximately constant in sequences
with diagenetic bedding (section 3), the somewhat variable shapes of
carbonate peaks, which can be found within single sections, must
primarily be a result of a fluctuating degree of mechanical compaction
(that is, the amount of compaction at the onset of lithification in
the middle of the limestone layers, see section 2.4). Commonly, the
mechanical compaction in the middle of the limestone layers varies
from about ±15 to ±20%, causing carbonate curves with similar shapes
but with different widths in their carbonate peaks and with
fluctuating maximum carbonate contents (see Fig. 46). Nevertheless,
the variation in mechanical compaction affects only carbonate-poor
sediment and, therefore, changes the maximum carbonate content in
those layers. As the compaction law shows (see Fig. 7), the influence
of compaction on carbonate content is relatively high at high absolute
clay contents when compaction is between 0 and 60%. This interval is
equivalent to the different amounts of mechanical compaction seen in
numerous alternations studied.

The different manners in which the fluctuating degree of
mechanical compaction acts upon the maximum carbonate content explains
the observations described in the literature concerning the
relationship between carbonate content and the thickness of limestone
layers. These data appeared contradictory, since limestone layers
with carbonate contents less than 35% show a direct relationship
between increasing carbonate content and increasing layer thickness.
However, such a relationship is not or only weakly found in limestone
layers containing more than 90% $CaCO_3$ (SEIBOLD, 1952; FLÜGEL &
FENNINGER, 1966; FLÜGEL, 1968).

As a result of the various carbonate curves presented in Fig. 46,
the theoretical relationship between the thickness of limestone layers
and maximum carbonate content for the Angles and Neuffen sections is
shown in Fig. 47. When compaction at the onset of lithification is
between 0 and 60%, a clear relationship between layer thickness and
maximum carbonate content exists only when the absolute clay content
is more than 5%. In accordance with that, measured data from the
Neuffen section (SEIBOLD, 1952) correspond well to the theoretical
curves (Fig. 48).

Based on the compaction law and the simplified assumption that
compaction steadily increases per interval of lithification (from the
middle of the limestone layer), the different types of carbonate peaks

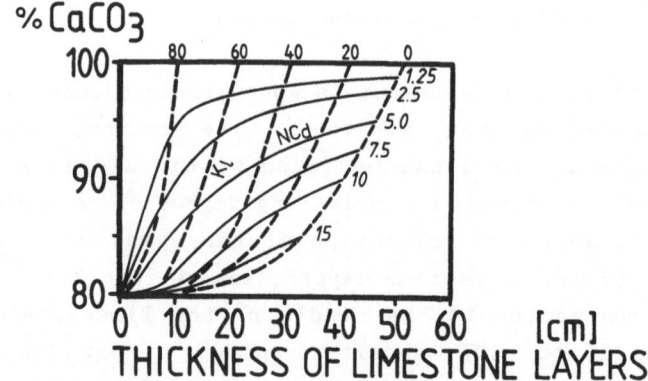

Fig. 47 Theoretical relationship between the carbonate content in the middle of the limestone layers and their thickness found in Jurassic and Cretaceous sections in southeastern France and southern Germany. The values are calculated from the carbonate curves presented in Fig. 46 and the formula for the carbonate content at the weathering boundary calculated in Fig. 43.

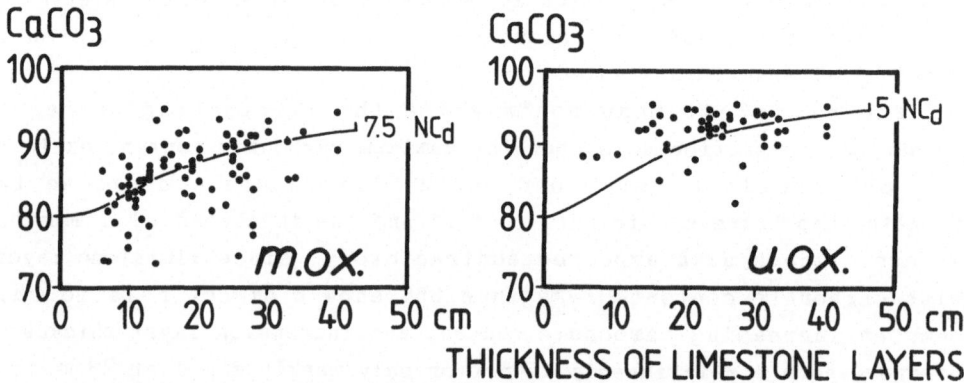

Fig. 48 Thickness of limestone layers and amount of carbonate in the Neuffen Quarry (from SEIBOLD, 1952). Data fit the theoretical curves from Fig. 47 with absolute clay contents of 7.5% (Middle Oxfordian) and 5% (Upper Oxfordian).

in marl-limestone alternations can be satisfactorily explained. Primary differences in composition produce the existing stratification only in so far as they trigger diagenetic bedding. The different shapes of the carbonate curves are mainly a result of different diagenetic histories which depend on the absolute clay content and the degree of the prediagenetic, mechanical compaction. Even limestone layers with extremely angular carbonate curves are not necessarily a product of an abrupt, violent change in the primary composition.

4.3 Development of Limestone Layers as
Related to Depositional Events

Event beds with an allochthonous composition, such as turbidites or tempestites, may cause special bedding phenomena because they often undergo preferred cementation. Most conspicuous of these are the so-called "underbeds" (MEISCHNER, 1964; EDER, 1971; 1982) which form because the cemented zone extends beyond the base of the event bed (see Fig. 49). Consequently, during diagenesis the event bed and several centimeters of the underlying sediment column become a single limestone layer. In principle, one also has to assume the existence of "upperbeds". However, they are difficult to recognize, for the most part, because the top of the event bed is often poorly defined due to post-event bioturbation and diminuation of grain size towards the upper contact.

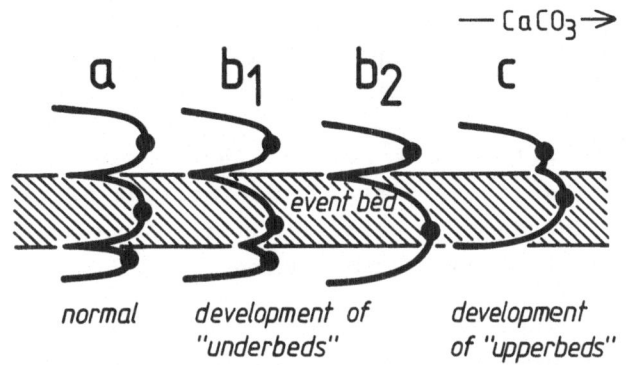

Fig. 49 Plots of carbonate variations typical for various lithified event deposits. Center of cementation (●).

The "underbeds" described in this study are products of burial diagenesis. Therefore, they are not to be confused with similar phenomenon of very early diagenetic cementation zones below event beds which have been secondarily burrowed and postdate early cementation (e.g., skeletal tempestites in the German Muschelkalk; AIGNER, 1982; 1985).

The following types of carbonate curves are formed by preferred cementation in depositional event beds (Fig. 49). The curves are determined somewhat by the location of the center of cementation, that is the site of the highest carbonate content and the lowest amount of compaction which is usually located in the middle of the limestone layer (see section 2.4).

1. The center of cementation is in the middle of the event bed; the lower and upper edges of this bed correspond to the decreasing proportions of the $CaCO_3$ curve (Fig. 49a). No "underbed" is formed.

2. The center of cementation shifts towards the coarse-grained, lower part of the event bed (Fig. 49b$_1$). The lower part of the layer is partly able to withstand mechanical compaction because of the grain-supported matrix or due to unstable carbonate phases which promote early cementation. However, if mechanical compaction is low, the remaining pore space, which is cemented, is high. For that reason, the carbonate content at the event bed base and several centimeters below is usually higher than the carbonate content of the weathering boundary. Consequently, an "underbed" is formed. In some cases, the center of cementation in the event bed and in the underlying bed can merge to form a single center of lithification (Fig. 49b$_2$). EDER (1971) described several event beds and "underbeds" displaying carbonate curves of the types b$_1$ and b$_2$.

3. "Upperbeds" supposedly form if the center of cementation is located in the fine-grained, upper portion of the event bed. There, for instance, a certain mineral composition might cause a relatively early onset of cementation.

In the following, two examples of the origin of carbonate curves types a and b are presented:

Calciturbidite of the Scaglia Rossa, Italy

Section: Quarries are located near Fossombrone, The Marches, Italy. The Red Scaglia Rossa limestones are interbedded with bleached, almost totally cemented turbidites about 30cm thick (LABUDE, 1983). One of those layers was in detail studied which exhibits an "underbed" locally and is sometimes separated from the underlying rock by one or several stylolitic bedding planes (Figs. 50, 51). It was possible to determine the degree of compaction because the entire layer is slightly affected by post-event bioturbation (mainly Chondrites).

Calculations and results: The turbidite layer displays its highest carbonate content (97%) in the middle of the event bed,

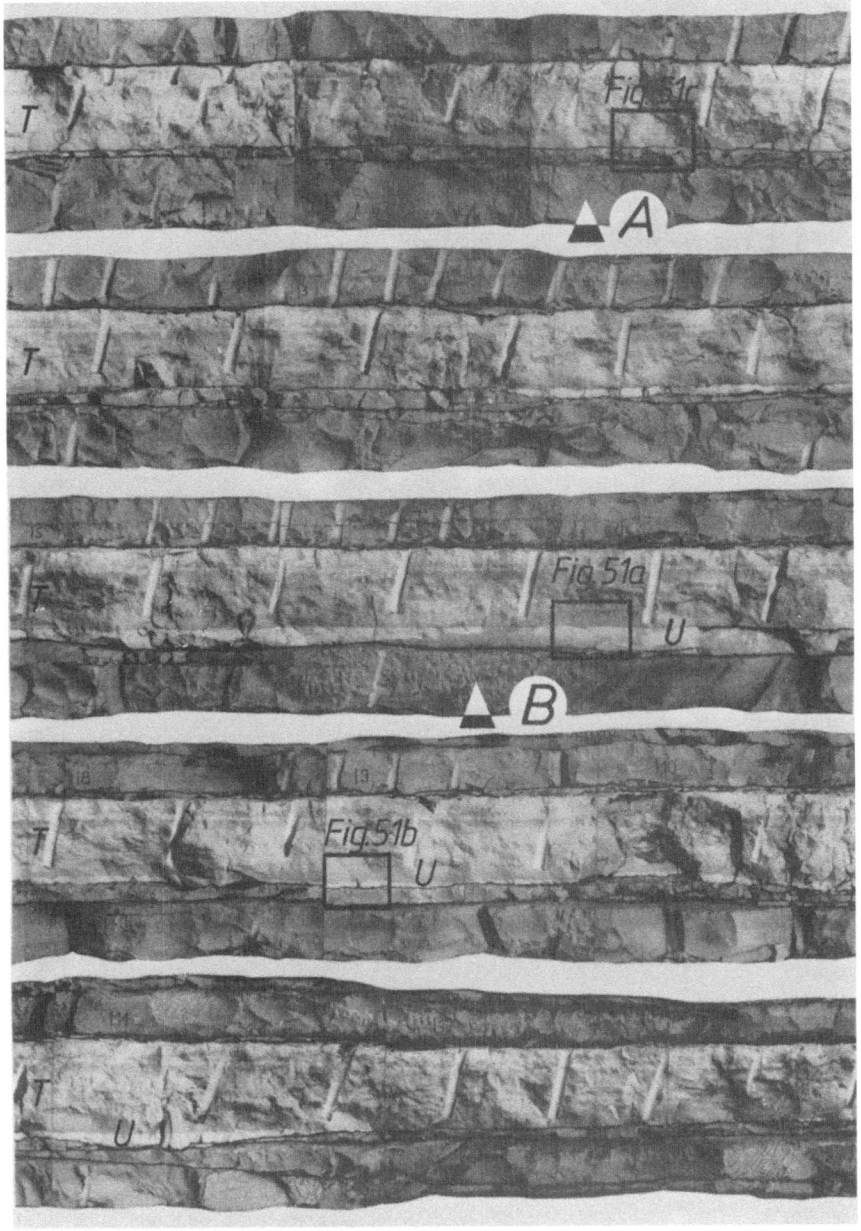

Fig. 50 Bedding phenomena due to deposition and diagenesis of a calcareous turbidite bed (light colored layer). The length of the strata shown is 15m. Note the appearance of a bleached "underbed" (U) at several sites below the turbidite bed (T). For sections A and B see Fig. 52. Maastrichtian Scaglia Rossa, Fossombrone, Italy.

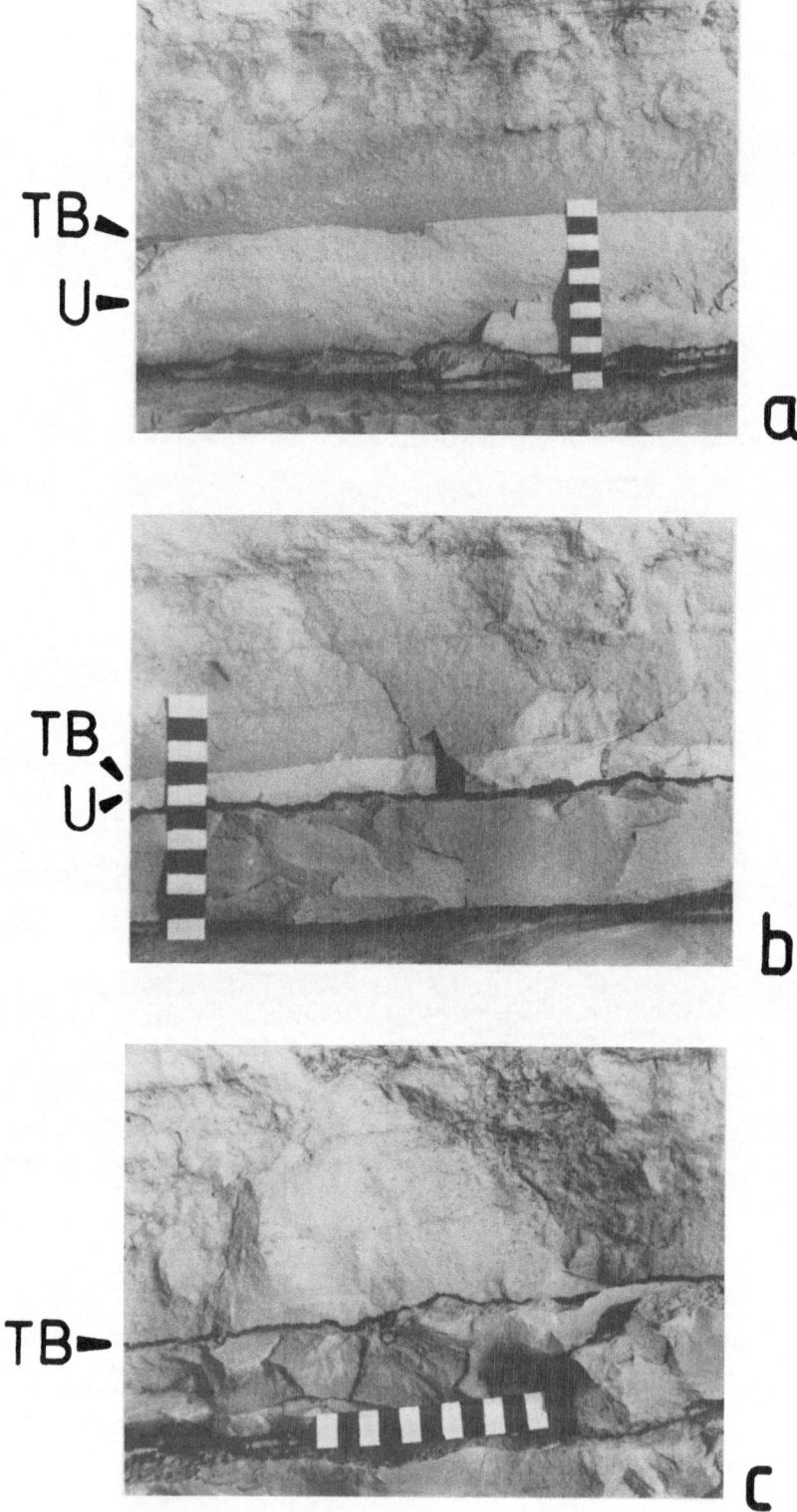

although its original carbonate content probably decreased slightly upwards as the absolute clay content increased from approximately 1-2% to 3-4%. After a relatively short phase of mechanical compaction (10 to 20%), lithification began early in the turbidite, since mechanical compaction was about 10 to 30% less compared to the surrounding sediment; additionally, the pore space became practically non-existent due to complete cementation in the event layer. Assuming that the behavior of porosity was similar to that known from recent deep-sea carbonates (HAMILTON, 1976), lithification of the turbidite began at an overburden between 40 to 80m, whereas lithification in the neighboring limestone layers began much later at an overburden between 150 and 300m. Probably, diagenetically unstable carbonate phases from the shallow sea (i.e., aragonite, Mg-calcite), which favored an early onset of cementation, were transported into the pelagic Scaglia Rossa (see SCHLAGER, 1980). The pelagic Scaglia Rossa was mainly composed of foraminiferal calcite ooze.

In Fig. 52A, the center of cementation (which is identical to the compaction minimum) lies in the upper part of the turbidite bed. Later, another, smaller center of cementation formed in the lower part of the turbidite. Its carbonate curve greatly decreases towards the turbidite base where a diagenetic marl joint developed. An "underbed" did not form.

In contrast, in Fig. 52B the center of cementation shifted slightly downward in the turbidite bed. Probably for that reason, the secondary center in the turbidite and the cementation center in the underlying limestone bed combined to form a single secondary center. This formed under about 50m more overburden as compared to the main center in the middle of the turbidite. However, lithification began much later around the bases of the turbidite between the first and second center of cementation because compaction increased to approximately 65%. Nevertheless, comparatively high compaction (65%) did not cause an appreciable decrease in carbonate content because the absolute clay content is very low (2 to 5%). This is evident from the carbonate compaction law (see Fig. 7). For that reason, the turbidite bed and the underbed are contained within a single limestone layer (see the carbonate curve in Fig. 52B).

Fig. 51 The base (TB) of the turbidite layer which is represented in Fig. 50.
a,b: Sharp, jointless contact with the lower limestone layer forms the "underbed" (U).
c: The turbidite bed is separated from the lower limestone layer by a stylolitic marl joint. Scale shown uses cm. Scaglia Rossa, Fossombrone, Italy.

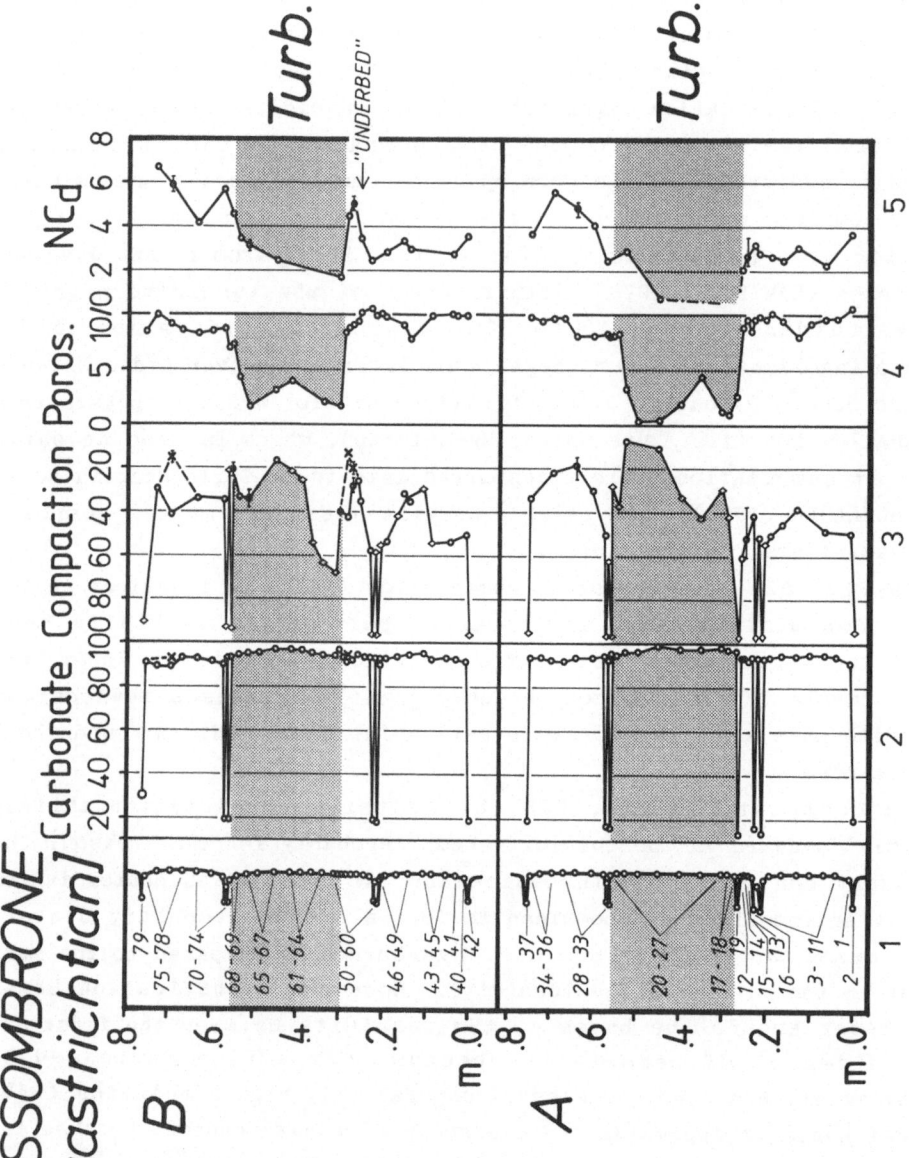

Fig. 52 Detailed study of the turbidite bed presented in Fig. 50 (shaded) both without (A) and with an "underbed" (B). Columns: 1) weathering profile and sample numbers, 2) carbonate content of selectively cemented burrows (x) and of the rock matrix (o), 3) compaction of selectively cemented burrows (x) and of the rock matrix (o), compaction is interpolated by using eq. 4 and mean absolute clay contents (◊), 4) porosity, 5) noncarbonate fraction, standardized as a percentage of the original sediment volume.

Turbidite to Tempestite Sequence Composed of Calcilutite, Upper Jurassic, Southern Germany

Another example concerning the diagenetic alteration of primary event bedding is the Upper Oxfordian turbidite to tempestite sequence in southern Germany (Fig. 53, see section 8.2.3). Centers of cementation, which are recognizable because they form maxima on the carbonate curve, developed within every turbidite bed, "underbed," and other types of limestones. The centers of cementation are independent of the primary carbonate curve and are now mostly in the middle of the event layers. Often, the decrease in carbonate content between the various event beds does not reach below the carbonate content at the weathering boundary, which is at a carbonate content of approximately 90%. For that reason, the existing limestone ledges usually contain several beds (mostly event beds).

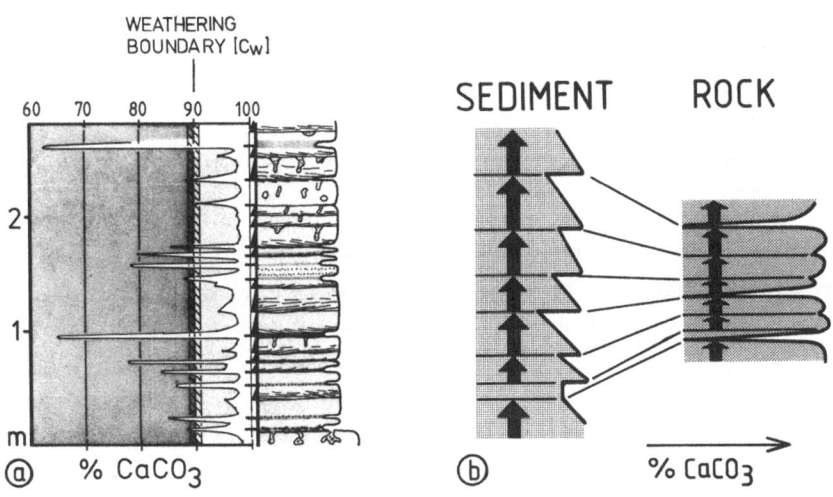

Fig. 53 Upper Oxfordian marl-limestone alternation, Geisingen Quarry, southern Germany.
a: Plot of variation in carbonate content with the carbonate content of the weathering boundary (C_W). The lithology column displays graded, partly laminated calcilutite beds which show post-event bioturbation (see Fig. 93).
b: Interpretation of the alternation as an diagenetically overprinted turbidite to tempestite sequence.

4.4 Diagenetic Enhancement of the Bedding Rhythm

Diagenesis modifies the amplitude, shape, and cycle frequency of the original carbonate curves. During diagenesis, rhythmic alternations form in which the thickness of the limestone layers is only slightly to moderately compacted, while that of the marl layers becomes considerably reduced due to intense compaction. The following sections deal with the various processes which affect the rhythmicity of diagenetic bedding (e.g., diminution of the number of primary carbonate cycles, decrease in the variations of maximum carbonate contents, and differential compaction between the dissolution and cementation zones).

4.4.1 Reduction in the Number of Primary Carbonate Oscillations

Not every primary carbonate oscillation is diagenetically enhanced, and not every enhanced carbonate oscillation weathers to form a limestone ledge. Therefore, diagenesis usually generates fewer limestone layers as compared to the number of primary carbonate oscillations. In certain circumstances, the reduction of the number of primary oscillations can amount to 40%. The following processes were observed:

1. Centers of cementation which are situated in close proximity to one another can result in a single major carbonate peak (see Fig. 1). Such layers are usually slightly thicker and have several smaller carbonate peaks than those with only one center of cementation. It is assumed that the centers of cementation are mainly located at those sites on the primary $CaCO_3$ curve which are maxima. Examples are shown in Figs. 20, 23, 25, 28, 32, 34, 37, 40, and 53.

2. Smaller carbonate variations with carbonate contents below that of the marl-limestone weathering boundary (EINSELE, 1982) weather completely to marl. This occurs especially in the marl-rich alternations (Type I, see Table 12). Examples are shown in Figs. 20 and 32.

   Primary carbonate oscillations can be detected at those sites where the post-diagenetic carbonate curve has maxima and minima (both for larger and smaller CaCO$_3$ fluctuations). Reduction in the number of primary variations for the carbonate-rich alternations (Types II and III) is between 5 and 20% (Table 13), primarily as a result of the processes just described in 1. However, reduction in the amount of primary cycles in marly alternations (Type I) is approximately 20 to 40%, due to the processes described in 1 and 2. If one considers only well-developed ledges, more than 60% of the primary carbonate oscillations can be lost during diagenesis and weathering.

Table 13: Diagenetic diminuation in the number of primary carbonate oscillations.

a Amount of primary carbonate oscillations in the studied sections (for symbols see Table 11)	b Amount of limestone layers with well and weakly developed ledges	c Amount of limestone layers with well developed ledges	b as percentage of a	c as percentage of a
R	33	28	17	84.8	51.5
A 1	24	18	9	75.0	37.5
A 2	18	15	15	83.3	83.3
A 3	15	13	12	86.7	80.0
L	10	9	9	90.0	90.0
N 1	60	37	22	61.7	36.7
N 2	28	25	22	89.3	78.6
G 1	46	42	36	91.3	78.3
G 2	46	44	22	95.7	47.8

4.4.2 Diminution of the Variations in Maximum Carbonate Content

The studied lithified sections (sections 3.2 to 3.5) provide clear evidence that carbonate-rich alternations have a nearly constant carbonate content in the middle of the limestone layers, whereas limestone layers in marl-rich sections display larger maximum carbonate variations (e.g., compare the carbonate curves of the sections Angles 1 and 3, 3.3.1 and 3.3.3, respectively). In order to explain this, one has to consider that fluctuations in mechanical compaction are commonly ±15 to ±20%. These variations cannot cause larger diagenetic carbonate differences when, in carbonate-rich alternations (Types II and III), mechanical compaction is 15 to 35% and the absolute clay content is low (<6%, see the compaction law, Fig. 7 and section 4.2). On the other hand, marl-rich alternations with high absolute clay contents (6 to 10%) undergo higher compaction (30 to 55%) until diagenesis sets in; hence, according to the compaction law, variations in the degree of mechanical compaction result in larger maximum carbonate variations.

The theoretical relationship between carbonate content and absolute clay content (described by the compaction law) is presented in Fig. 54 for different degrees of compaction. The diagram also gives the amount of $CaCO_3$ variation in both limestone layers and marl beds in each section studied. As expected, carbonate-rich sections have small differences in their maximum carbonate content as compared to marl-rich sections. In accordance with the compaction law, marl beds exhibit carbonate variations of more than 20%, although

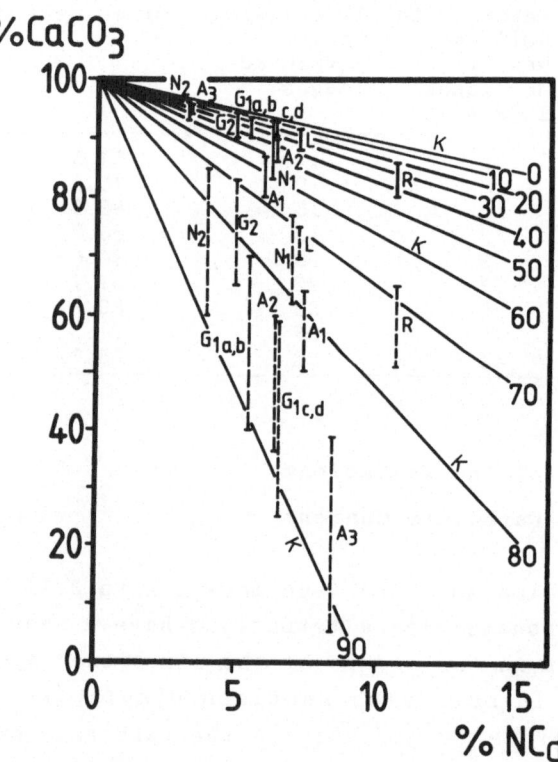

Fig. 54 Theoretical relationship between carbonate content, the standardized noncarbonate content (NC_d), and compaction (K=0 to 90%), calculated by using the compaction law. It was assumed that the porosity in lithified marl limestone alternations is 5%. Measured carbonate fluctuations in the various lithified alternations studied is shown for the middle of limestone layers (⊢——⊣) and marl beds (⊢--⊣).

compaction varies to the same degree as or to even a lesser amount than in the limestone layers; however, the total compaction is as much as 70 to 90% (see Fig. 7).

4.4.3 Differential Compaction Between Dissolution and Cementation Zones

An important result of differential compaction between the slightly compacted limestone layers (cementation zones) and the strongly compacted marl beds (dissolution zones) is that the original bed thickness changes to the detriment of the marl beds. Differential compaction is high when mechanical compaction in the limestone layers is low. Mechanical compaction depends in turn on the primary sediment composition. It is low when the original carbonate content is high (see Table 12 and Fig. 62).

Fig. 55 displays the variability in thickness reduction between dissolution and cementation zones calculated from the marl-limestone alternations studied. The reduction is expressed as the ratio of the compacted rock volumes (100-K) in the dissolution and cementation

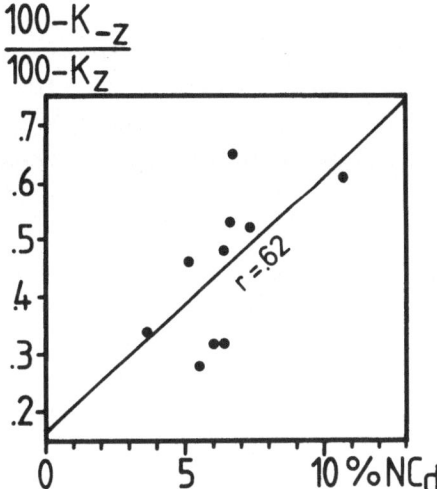

Fig. 55 Ratio of the thickness of dissolution zones to the thickness of cementation zones in various lithified marl-limestone alternations (vertical axis) versus the normalized noncarbonate fraction (NC_d, horizontal axis). The relative thickness of both zones is expressed as the difference between original thickness (100%) and the degree of compaction (%K).

zones by using average compaction values. Dissolution zones in
carbonate-rich alternations are reduced from one third to one quarter
of the thickness of the cementation zones, whereas in marl-rich
alternations the thickness of dissolution zones is only one half the
thickness of the cementation zones. For simplification, in this
calculation it is assumed that the original sediment had the same
thickness for both zones.

Differential compaction markedly affects the rhythmicity of
diagenetic bedding. Alternations with low mechanical compaction
(Types II and III; Table 12) have strongly reduced marl layers and are
dominated by limestone layers which have approximately equal maximum
carbonate contents and nearly the same thickness (sections 4.2 and
4.4.2). Therefore, these alternations now appear (after diagenesis)
to be more rhythmical or cyclical than as they were expressed in the
original sediment. However, even marl-rich alternations display an
enhancement of their original bedding rhythmicity due to differential
compaction, although enhancement is less than for carbonate-rich
alternations.

4.4.4 Relative Enrichment of Particles due to
Differential Compaction

Marl beds often contain a greater amount of particles with low
solubility as compared to the neighboring limestone layers. These
particles are composed of siliclastic and calcarenitic grains, fossil
detritus, agglutinated foraminiferas, and palynomorphs (e.g., WEBER,
1951; SEIBOLD et al., 1953; 1959; HALLAM, 1964; EDER, 1982). Although
this phenomenon might be in part due to primary effects, the particle
enrichment in the marl beds can also be easily explained by
differential compaction. It considerably reduces the volume of the
marl beds relative to that of the limestone layers (section 4.4.3).
If one assumes a mainly constant particle content in the primary
sediment, low soluble particles must be relatively enriched in the
highly compacted volume of the marl beds.

The amount of particles with low solubility (N) in a given volume
of rock or sediment depends on the initial particle concentration (N_O)
and the degree of compaction (K).

$$N = \frac{N_O}{(1-0.01K)} \quad . \tag{12}$$

If the enrichment is expressed as the ratio of particle concentration in neighboring marl and limestone beds (e), one then gets from eq. 12:

$$e = \frac{N_M}{N_L} = \frac{100-K_L}{100-K_M} \quad , \tag{13}$$

where the subscripts M and L indicate compaction in the middle of the marl and limestone beds, respectively. If one uses the average amount of compaction in the middle of several limestone layers (instead of K_L), this calculation would approximate the compaction at the onset of lithification (K_1).

The theoretical, mean factors of enrichment, which are calculated from the lithification compaction and the mean compaction of the marl layers, are between 2 and 5. The factor of enrichment increases if the lithification compaction decreases (Table 14). According to the enrichment factors, differential compaction in skeletal limestones should also affect their petrographical pattern. For instance, mudstones or wackestones in the limestone layers should convert to packstones in the marl beds as a result of the increasing compaction. This can be observed in several bedded Jurassic reef talus deposits in southeastern Germany (B. LANG, Erlangen; oral communication).

Table 14: Theoretical enrichment of nonsoluble particles in the marl beds relative to the limestone layer.

Section*	R	A1	A2	A3	L	N1	N2	G1A,B	G1C,D	G2
Compaction at the onset of lithific. (%)	33	55	36	26	28	27	19	33	28	27
Mean compaction in the marl beds (%)	69	81	81.2	82	73.2	71.8	82.1	85.6	85.2	78.7
Mean factor of enrichment (e)	2.2	2.4	3.4	4.1	2.7	2.6	4.5	4.7	4.9	3.4

*for symbols see Table 11.

4.5   Simulation of Diagenetic Bedding Rhythms

In the previous sections (4.1 through 4.4), bedding and diagenetic
rhythms were quantified such that the input of theoretical parameters
into three models could simulate the cyclicity seen in marl-limestone
sequences. In these simulation models, a calcareous ooze with an
initial composition of 72% $CaCO_3$ and 75% porosity was assumed; this
corresponds to an absolute clay content of 7%. The carbonate neutral
value between dissolution and cementation zones (C_n=72%) is equivalent
to the primary carbonate content, because the diagenetic carbonate
system was assumed to be closed. In the three models given below (1
to 3), the distance between two carbonate peaks was evaluated as
follows: First, the carbonate curve of a limestone layer was
evaluated (eqs. 5, 11) and then the cement content of the cementation
zone was determined (eq. 17, section 5.3). Thereafter, the thickness
of the dissolution zone was chosen so that the dissolved carbonate was
equivalent to half the cement content of the two neighboring limestone
layers. Rock porosity was neglected in the calculations.

1.   In the first model, it is assumed that compaction at the onset of
     lithification and compaction in the marl layers should always be
     constant at 55 and 85%, respectively. The data result in a
     constant carbonate content of 53.5% in the marl beds and in a
     constant carbonate content in the middle of the limestone layers
     of 84.4%. The resulting alternation has a perfectly regular,
     periodic oscillating carbonate curve which is not documented in
     the rock record and is therefore not presented here.

2.   However, in the second model (Fig. 56a) it is assumed that
     compaction at the onset of lithification (K_l) should vary randomly
     according to a normal distribution with a mean of K_l=55% and a
     standard deviation of 15%. This type of alternation is also
     poorly documented in the rock record, because the carbonate
     content in the marl beds remains constant, and carbonate peaks do
     not have several maxima.

3.   The third model (Fig. 56b) is a variation of model 2 where the
     compaction at the onset of lithification should vary randomly in
     the same way as in model 2. In addition, however, the original
     distance between two given centers of cementation should be
     randomly distributed within an interval of 2m. Calculations
     reveal that for this model comparatively large carbonate

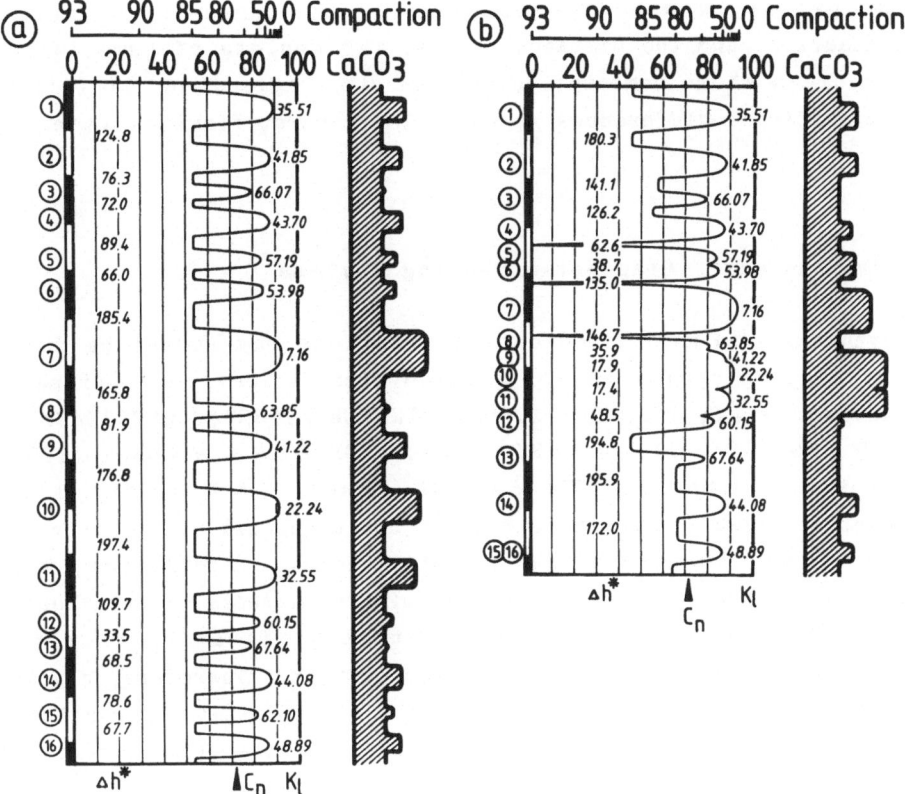

Fig. 56 Simulation models of marl-limestone alternations with the carbonate content curve (left) and the weathering profile (right, hatched). Scale is in meters. Values shown represent the interval (in cm) between the adjacent centers of lithification (Δh^*), and the compaction at the onset of cementation (K_1). C_n=CaCO$_3$ neutral value.
a: Compaction at the onset of lithification scatters randomly around a mean of K_1=55% by using a normal distribution curve with σ=15%. Compaction in the marl layers is constantly 85%.
b: Compaction at the onset of lithification as in Fig. 56a but with a random variation of 0 to 2m between the adjacent centers of lithification.

variations in the marl beds are produced. Moreover, limestone layers contain several carbonate peaks, as is the case in the rock record. In the weathered section, a typical "bundling" of limestone layers is caused which is comparable to some parts in the Neuffen 1 section (see Fig. 32). Model 3 shows that the diagenetic rhythmicity depends on the absolute clay content (or the primary sediment composition), the degree of mechanical compaction, and the distance between the centers of cementation. The sites where cementation sets in are triggered by the (slight) compositional differences in the primary stratification (see Fig. 1).

4.6   Conclusions: Diagenetic Bedding Rhythms

The rhythmicity in marl-limestone alternations may indeed be due to primary bedding cycles (e.g., Milankovitch cycles), as described by many authors, e.g., FISCHER (1980); EINSELE (1982), SCHWARZACHER & FISCHER (1982); COTILLON & RIO (1984); DE BOER & WONDERS (1984); FISCHER et al. (1985), BOTTJER & ARTHUR (1985); DEAN & GRADNER (1985); BARRON et al. (1985).   However, this is not necessarily a requirement. The results obtained show that the diagenetic overprinting, which considerably enhances the bedding rhythmicity, affects primary bedding cycles as well as sequences with event stratification or with a randomly, slightly oscillating carbonate content.   Therefore, rhythmic carbonate oscillations in the present rock offer little information on the bedding pattern (cyclic or not cyclic) in the primary sediment.

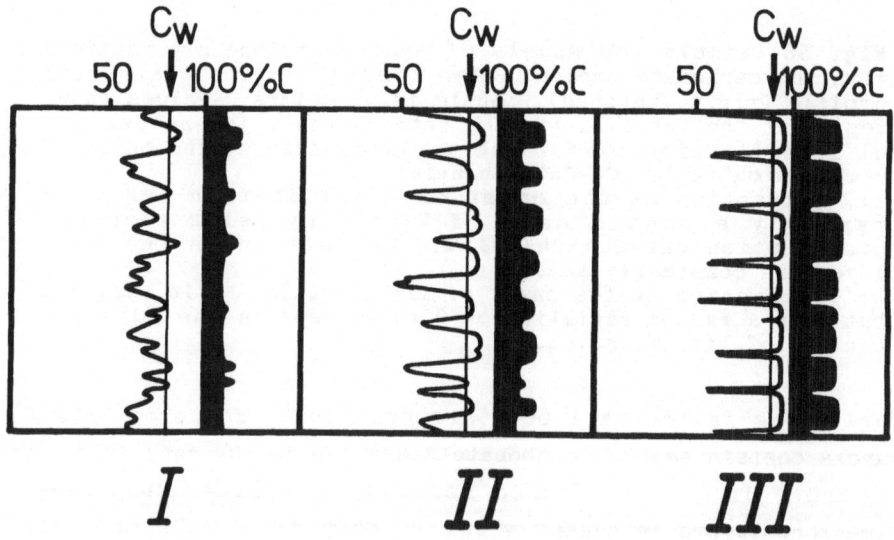

Fig. 57 Types of lithified marl-limestone alternations, carbonate curves, and weathering profiles (black). C_w=carbonate content of the marl-limestone weathering boundary, C=CaCO$_3$ content in weight percent.

Also, the (higher) number of initial carbonate oscillations cannot be determined with certainty from the present number of limestone layers.

Assuming that the carbonate content at the weathering boundary is constant, the development of the different types of diagenetic bedding depends solely on three parameters. These are the distance between the centers of lithification, which is controlled by the primary stratification, and two interrelated parameters, the absolute clay content and the compaction at the onset of lithification. Compaction at the onset of lithification (that is mechanical compaction in the middle of the limestone layers) is greater when the absolute clay content is higher. Consequently, qualitative results concerning the primary sediment composition and its diagenetic history can be obtained from the post-diagenetic bedding rhythms and the shapes of their carbonate peaks (Fig. 57).

Types of Bedding Rhythms in Fine-Grained, Lithified Carbonates

Type I   Marly Alternations

According to the carbonate compaction law, the alternations have sinusoidal carbonate curves due to high mechanical compaction (35 to 55%) and high absolute clay content (6 to 10%). Moreover, the maximum carbonate content in several limestone layers must vary widely because fluctuations in the degree of mechanical compaction affect the carbonate content. Therefore, and because of the relatively small amount of differential compaction, diagenetic bedding produces only weak rhythmicity. The maximum thickness reduction of the dissolution zones is approximately one half of the thickness of the cementation zones. The rhythmicity is further weakened because lower maxima on the carbonate curve are below the $CaCO_3$ weathering boundary; hence, these maxima appear as marl beds in the outcrop.

Type II   Alternations with Medium to High Carbonate Content

Both mechanical compaction in the limestone layers and the absolute clay content have intermediate values (25 to 35% and 4 to 6%,

respectively). Therefore, the carbonate compaction law gives
convex-shaped carbonate curves, in which the maximum carbonate content
of several limestone layers varies little. Moreover, differential
compaction between dissolution and cementation zones reduces the
thickness of the dissolution zones (marl beds) to about one half or
one third the thickness of the cementation zones (limestone layers).
Reduction of diagenetic carbonate cycles due to the influence of
weathering is medium to low.

Type III Highly Calcareous Alternations

Mechanical compaction as well as the absolute clay content are low
(less than 25% and less than 4%, respectively). Thus, differential
compaction is intense. The thicknesses of the marl beds is reduced to
more than one third the thickness of the limestone layers; and,
therefore, the rhythmicity is great. Weathered sections display
brick-like alternations, as described above, where limestone ledges
have a relative constant thickness. According to the compaction law,
limestone layers must have angular carbonate curves, while the
carbonate content is constant in the middle of the limestone layers.

5 DIAGENETIC BEDDING: CAUSES AND SIMULATION MODELS

5.1 Causes and Processes of Burial Carbonate Redistribution

In the sections studied, diagenetic bedding, which predominantly is a stratiform carbonate redistribution process, begins when overburden reaches thicknesses between 80 and 460m (see Table 12). Presumably, carbonate redistribution is not caused by chemical peculiarities between primary beds which had slight differences in carbonate content. This is evident from the calculated original composition, which indicates that primary mean carbonate differences between dissolution and cementation zones were only small (on an average 5.8%). Moreover, carbonate reprecipitation occurs even in the pore space of the marl beds to form typical pressure shadow structures (Figs. 30F,G; 58) when the lithostatic stress is laterally reduced around massive fossils which are difficult to compact (e.g., belemnites).

Pressure shadow structures clearly have a higher carbonate content as compared to the surrounding marl beds, although their carbonate content is not as high as that in the adjacent limestone layers. Probably, this is caused by primary compositional differences and by a greater pre-diagenetic compaction of the pore space in the marl beds as compared to that in the limestone layers (Table 15). Pressure shadow structures are especially well developed in carbonate-rich alternations which underwent low levels of mechanical compaction and which were accompanied by a relatively early onset of diagenesis (Type III). On the other hand, in alternations with high levels of mechanical compaction and usually a late beginning of diagenesis (Type I), the pore space adjacent to massive fossils is already too restricted to receive significant amounts of carbonate cement.

The occurrence of both stylolitic, flasery marl seams as it is presented in Fig. 30H (WANLESS, 1979; TRURNIT & AMSTUTZ, 1979; RICKEN & HEMLEBEN, 1982) and pressure shadow structures within the same marl bed indicates that the carbonate redistribution must be related to differences in stress (e.g., WEYL, 1959; NEUGEBAUER, 1974; ROBIN, 1978; ENGELDER et al., 1981). In the marl bed, pressure dissolution of CaCO$_3$ is presumably caused by lithostatic stress which acts upon uncemented grain contacts. The dissolved carbonate is reprecipitated in the marl beds at those sites where the lithostatic stress decreases locally in response to the presence of solid pressure conductors (i.e., heavy shells, Fig. 58).

PRESSURE SHADOW STRUCTURES

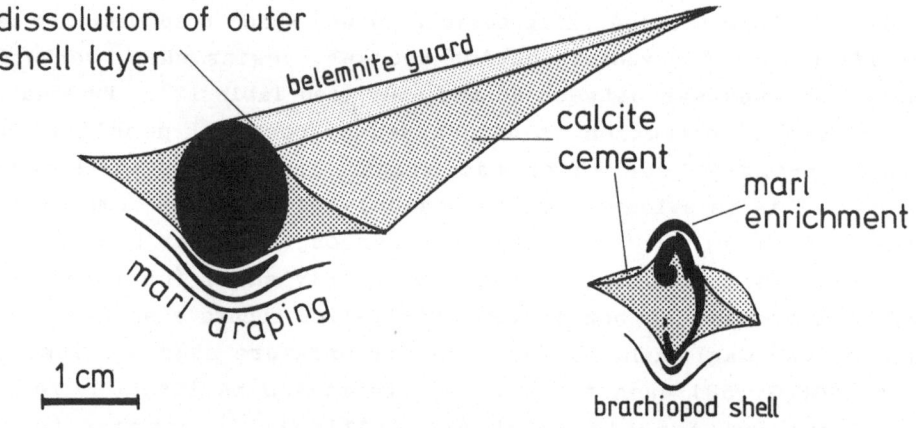

Fig. 58 In calcareous pressure shadow structures, carbonate precipitation occurs laterally of slightly compacted to uncompacted fossils in response to the diminution of the lithostatic stress. Calcareous pressure shadow structures are frequently found in the marl layers of the marl-limestone alternation Type III.

The formation of calcareous pressure shadow structures may serve as a model for the carbonate cementation in limestones. In limestone layers and pressure shadow structures, the original carbonate content is greatly enhanced due to cementation; however, only pressure shadow structures provide a clear insight into the hypothesis that carbonate cementation during burial diagenesis may be caused by differential stress between the layers of a bedded sequence. During lithification, the marl beds should have a higher lithostatic stress at the grain contacts than the limestone layers. As CORRENS (1949), NEUGEBAUER (1973; 1974) and WALTER & MORSE (1984) demonstrated, the intensity of the pressure dissolution-reprecipitation process depends on the specific lithostatic pressure, the mode and number of grain contacts

Table 15:   Carbonate content and the degree of compaction in pressure shadow structures (PSS) lateral of belemnite guards.

Type of PSS	Mean carbonate content in PSS	Carbonate content in the adjacent marl bed	Standard-ized non-carbonate content in the marl bed	Compaction in the PSS	Compaction in the adjacent marl bed*	Mean compaction at onset lithif. in limestone layers
	82.2	44.3	8.5	48.9	81.8	26.4
A3	86.2	42.5	8.5	35.2	82.4 75-83**	26.4
◁○▷	88.4	7.8	8.5	23.3	88.7	26.4
	86.7***	24.6	8.5	32.8	86.3	26.4
N2	83.8	77.2	4.0	74.0	81.2	19.3
◁○▷	84.8	74.8	4.0	72.6	82.9	19.3

*calculated by using eq. 4 and the regression curves between porosity and carbonate content as described in Figs. 27 a and 35 a.
**measurement.
***carbonate reprecipitation within the belemnite phragmocone.

and grain size, the mineral phases involved, and the composition of the pore solution.

One can expect that the dissolution-reprecipitation process, which obviously seems to be triggered by stress differences in the grain structure of marl and limestone layers, is self-perpetuating. During the phase of mechanical compaction, the lithostatic pressure at the grain contacts should be relatively homogeneous at first. However, when diagenetic carbonate redistribution begins, differential stress developed in the sediment. At the grain contacts in the limestone layers, the stress continuously decreases due to the welding of grains by carbonate precipitation. Whereas, in the marl layers, the pressure affects the unwelded grain contacts, resulting in carbonate dissolution. In analogy with the above-mentioned $CaCO_3$ pressure shadow structures, the dissoloved carbonate is reprecipitated at sites with low lithostatic pressure (WEYL, 1959), i.e., in the later limestone layers. This process is self-perpetuating since the reprecipitated carbonate causes a further pressure decrease in the grain structure, which in turn favors cementation. The self-perpetuating, dissolution-reprecipitation process probably comes to a standstill when an equilibrium is reached between pressure dissolution in the marl beds and opposing factors such as decreasing permeability, the increasing amount of insoluble residue, and the increasing density of packing in the marl layers. Usually, dissolution-reprecipitation ceases before the pore space in the limestone layer is totally filled (the porosity in limestone layers is between 0 and 5%).

Obviously, stress differences within the grain structure between marl and limestone layers are triggered by the <u>first, slight cementation in the relatively carbonate-rich beds</u>. This can be explained as follows:

1. The higher the primary carbonate content, the greater is the probability for the $CaCO_3$ dissolution-reprecipitation process to be initiated. Moreover, the various beds of a primary sequence can contain different amounts of unstable carbonate phases. Both factors determine the "diagenetic potential" of an original sediment (SCHLANGER & DOUGLAS, 1974). On the other hand, the presence of clay minerals seems to hinder the reaction between the calcareous phases (BAKER et al., 1980) and to hamper the formation of larger calcite crystals during cementation (MARSCHNER, 1968; BAUSCH, 1968). Moreover, the exchange of trace elements which contain in the carbonate fraction (e.g., Sr) is hampered by an increasing clay content (see section 7.2).

2. Differences in the compaction of primary beds relatively richer and poorer in carbonate content could trigger diagenetic bedding. Usually, clays undergo a higher degree of compaction as compared to pelagic carbonates (see Figs. 60, 61). Given a primary sediment containing slight carbonate oscillations, the porosity in the carbonate-poor layers is somewhat reduced due to higher compaction relative to that in the carbonate-rich layers (see Fig. 15). Thus, in the marl beds the comparatively large carbonate grains are subjected to more stress than the smaller clay minerals. (The diameters of coccoliths and clay minerals are roughly 5 and $1-2\mu m$, respectively.)

The assumption that differential stress between marl and limestone layers triggers diagenetic bedding is supported by results from the Deep Sea Drilling Project. In the ooze/chalk transition zone of several sections some layers harden while others remain soft (SCHLANGER & DOUGLAS, 1974; GARRISON, 1981). MATTER (1974) examined calcareous sequences from the Arabian Sea (60, 70, and 200m thick) consisting of alternations ranging from soft and semilithified ooze and chalk layers (sites 220 and 223). The different degrees of cementation in these sequences reflect slight differences in the primary composition. In agreement with the results obtained in this study, the onset of cementation occurs at a porosity of approximately 60% and at an overburden of 100 to 200m. Presumably, larger stress differences could arise to trigger diagenetic bedding if these sequences are subjected to greater overburden.

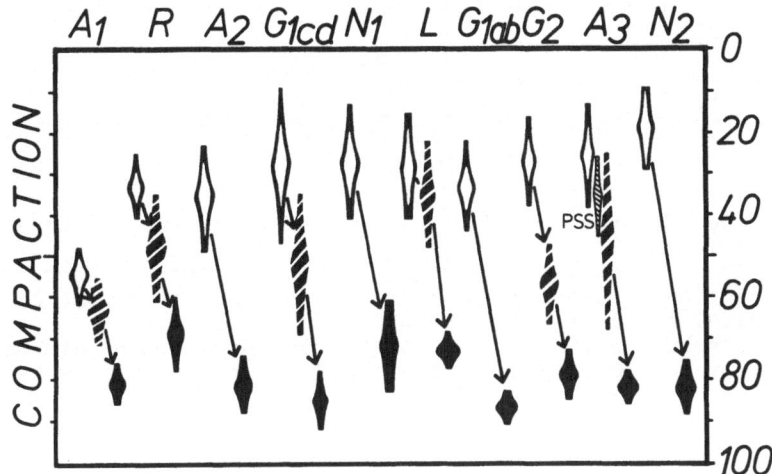

Fig. 59 Compaction in marl-limestone alternations. Shown are means and areas of scatter for compaction. White: compaction in the middle of the limestone layers (that is, compaction at the onset of lithification), hatched: compaction of selectively cemented burrows, black: compaction in the middle of the marl layers, PSS: compaction in calcareous pressure shadow structures.

In this study, diagenetic carbonate oscillations are always related to differences in compaction. Simply stated, differences in compaction between marl and limestone layers are greater when mechanical compaction in the middle of the limestone layers is lower. This is because compaction in the marl layers is always approximately the same (about 80%), whereas mechanical compaction in the various sequences is different depending on the primary carbonate content (Fig. 59). As already reported by MATTER (1974) and SCHLANGER & DOUGLAS (1974), mechanical compaction is higher when the primary carbonate content of the sediment is lower (Fig. 60a). Carbonate-rich and marly sediments undergo an average mechanical compaction of 20 and 40%, respectively. According to the behavior of porosity in recent

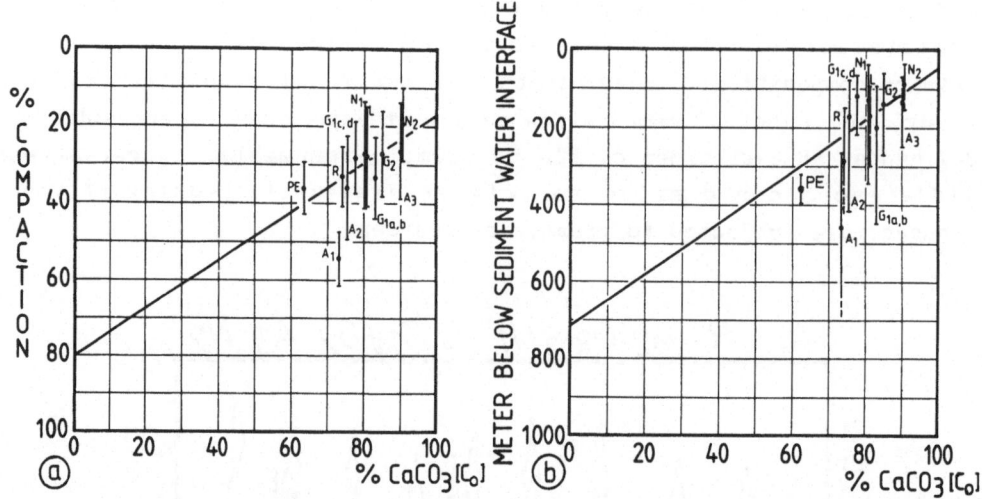

Fig. 60 Compaction and meters of overburden at the onset of lithification.
a: Compaction at the onset of lithification (or mechanical compaction) versus the mean primary carbonate content in the original sediment of the cementation zones.
b: Overburden (in meters) versus the mean primary carbonate content in the original sediment of the cementation zones.

pelagic carbonates (section 2.4), this is equivalent to 100 and 400m of overburden (Fig. 60b).

In the middle of what later becomes the limestone layer, the burrow system and the surrounding sediment (after cementation) usually have the same carbonate content. Therefore, during the onset of cementation, the process occurs predominantly simultaneously in the burrows and the sediment matrix at about the same degree of compaction. However, when lithification proceeds towards the outer edges of what are later the limestone layers, the precipitation of carbonate cement is quite selective. It often occurs earlier in burrow systems and fossils, whereas the neighboring sediment is subjected to further compaction (Fig. 59, see Fig. 4c). Despite this, mechanical compaction and thus the time required for the onset of lithification increase from the middle of the limestone layers towards the marl layers in both the burrow system and in the surrounding sediment (for examples see Figs. 18, 28). Obviously, the microchemical environment in burrows and fossils is suitable for carbonate cementation, supposedly because they often have a higher organic content, a higher pH in the interstitial solutions, and a higher porosity.

The transportation of the carbonate released from the marl beds must occur with and against the flow direction of the expelled pore fluid. In the Upper Jurassic of southern Germany, carbonate from dissolution-affected bedding planes is reprecipitated both upsection and downsection in the two neighboring limestone layers (see Fig. 9, section 2.3.1). In addition, limestone layers in the alternations studied mainly have symmetrical $CaCO_3$ curves, indicating carbonate migration from both above and below. Thus, during burial diagenesis transport by diffusion (BERNER, 1980; PINGITORE, 1982) must be much more effective than transport due to upwardly moving pore waters (EINSELE, 1977; WEDEPOHL, 1979; BATHURST, 1980b).

Further evidence for this is the unrealistically large amount of time needed for the cementation of limestone layers if their carbonate cement was exclusively transported by the expelled pore waters. In order to cement a typical, 30cm thick limestone layer (with 90% $CaCO_3$ of which 40% is cement), one needs an 11cm thick layer composed of pure $CaCO_3$ cement. A column 11cm by $1cm^2$ from this layer is equivalent to 30g $CaCO_3$ cement. If cementation begins at an overburden of 200m and at 60% porosity (which is commonly found in marl-limestone alternations), the velocity of the rising pore water would be, as EINSELE (1977) shows, about half the sedimentation rate. Assuming that all of the dissolved carbonate in the pore water is consumed during cementation (which, however, is not very realistic; see BATHURST, 1976) maximum carbonate precipitation would be $1.9mg/cm^2$ in 1000 years. This is based on sedimentation rates as high as 0.05m/1000 years and commonly found concentrations of Ca in the pore water of 0.5g/l (NEUGEBAUER, 1974; SALES & MANHEIM, 1975; GIESKES, 1984). These calculations result in an unrealistic, minimum time of 15 million years which is necessary to solely cement the above-mentioned limestone layer by upmoving pore waters.

Minimum time intervals in the order of 100 million years are necessary in later stages of diagenesis, since the velocity of the rising pore fluid decreases rapidly with increasing overburden. For example, with 500m of overburden, the velocity of the pore water is on the order of only one tenth of the sedimentation rate (EINSELE, 1977).

5.2 Burial Reduction and the "Inversion" of Porosity

During the deposition of an overburden, clay and calcareous ooze display a different porosity reduction behavior (Fig. 61, data

compiled from BALDWIN, 1971; HAMILTON, 1976; SCHOLLE, 1977; LOCKRIDGE
& SCHOLLE, 1978; BALDWIN & BUTLER, 1985). Originally, clays have a
relatively large reduction in porosity, but with 500 to 1000 meters of
overburden, porosity reduction becomes low. In contrast, calcareous
ooze has a more linear decrease in porosity; therefore, the two curves
intersect. At certain amounts of overburden, sequences containing
$CaCO_3$ variations must show a different relationship between porosity
and carbonate content.

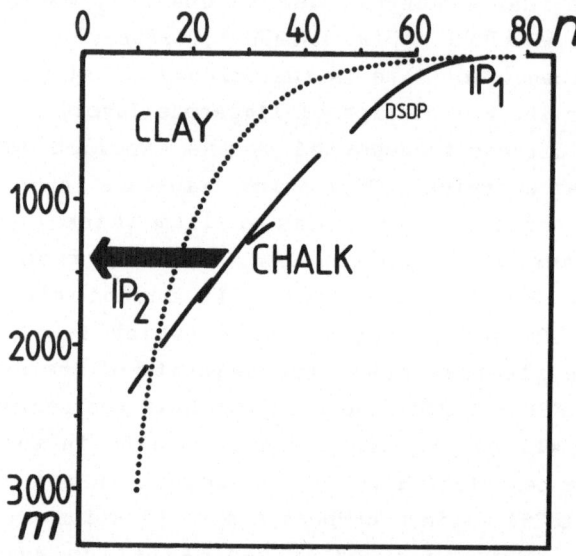

Fig. 61 Porosity versus overburden in fine-grained
carbonates and clays. $IP_{1,2}$: "inversions" of porosity;
arrow: porosity reduction in chalks due to cementation.
Curves compiled from BALDWIN, 1971; HAMILTON, 1976; SCHOLLE,
1977; LOCKRIDGE & SCHOLLE, 1978; BALDWIN & BUTLER, 1985.

The first "inversion" of porosity occurs during the stage of
mechanical compaction with 10 to 100m of overburden. Although the
porosity is originally somewhat higher in carbonate-poor beds (section
2.3.2), porosity is reduced in those beds by a relatively higher
amount of mechanical compaction. Thus, porosity increases with
increasing carbonate content (see Fig. 17).

The second "inversion" of porosity takes place with several 100's
or 1000's m of overburden, due to the intersection of the two
porosity-overburden curves (Fig. 61, P_2) or the cementation of
calcareous beds. Cementation of the carbonate-rich layers nearly
fills their pore space, whereas the porosity of the carbonate-poor

layers must progressively follow the porosity-overburden curve for clay due to the increasing impoverishment in carbonate. Therefore, when carbonate redistribution ceases, the porosity in nearly all observed lithified marl-limestone alternations increases with decreasing carbonate content in the same manner (Fig. 62). If one extrapolates the porosity of the marl beds to carbonate-free rock, it yields porosities of 15 to 25%. According to the porosity-overburden curve for clay (BALDWIN, 1971; BALDWIN & BUTLER, 1985), an overburden of 500 to 1500m is necessary corresponding in its order of magnitude to the observed amount of overburden (see Table 11).

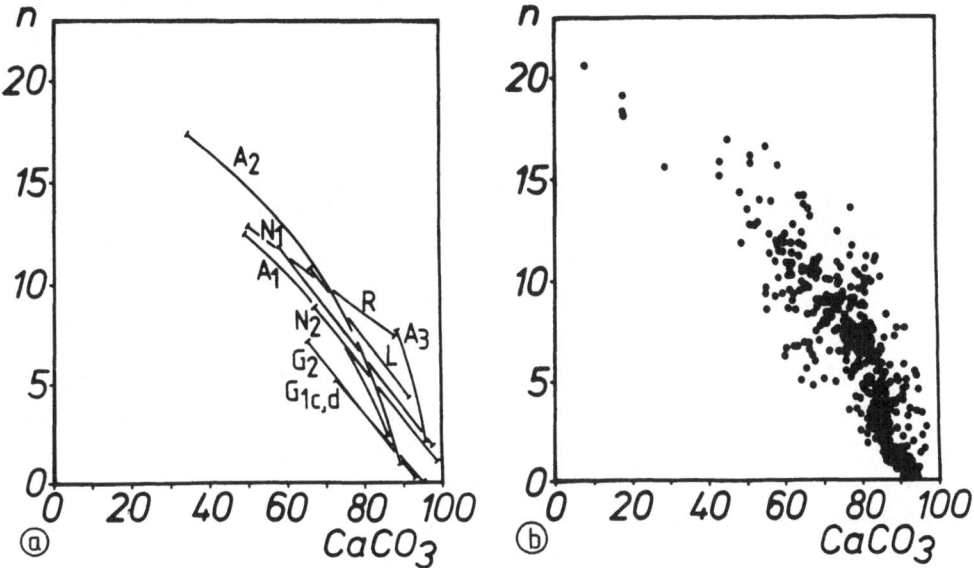

Fig. 62 Relationship between porosity (n) and carbonate content (C) in lithified marl-limestone alternations.
a: Regression curves of various marl-limestone alternations.
b: Combined plot of all porosity and carbonate data.

5.3 Calculation of Cement Content and

Primary Carbonate Content

One of the most promising means for quantifying the diagenetic mass exchange is the use of formulas which calculate the amount of $CaCO_3$ cement. In this study, the numerical determination of the cement content serves as an instrument in calculating and simulating the process of diagenetic bedding. The term "cement" is here used in a somewhat broad sense. Positive values of cement in a given rock sample indicate the amount of diagenetically precipitated carbonate, whereas negative values indicate the amount of dissolved carbonate. In the following discussion, the cement content (Z) is expressed in three different forms:

1. The "absolute cement content" (Z_d) which is normalized to the decompacted primary sediment volume (expressed in vol.%).

2. The "relative cement content" (Z_c) which is normalized to the total carbonate fraction. It is most accurately expressed as volume percent. However, the relative cement content is practically equivalent to the amount expressed as weight percent, because the specific densities of the carbonate and noncarbonate fractions are commonly very similar.

3. The "cement number" (z); this is a dimensionless ratio between the absolute cement amount (Z_d) and the primary, absolute carbonate content (C_{od}).

The carbonate compaction law alone is not sufficient to determine the cement content (see section 2.2) because no distinction is made between the primary and the cemented amount of $CaCO_3$. Therefore, either the primary porosity (n_o) or the primary carbonate content (C_o) must be added to the four parameters related in the compaction law (these are compaction, K; carbonate content, C; porosity, n; and the standardized noncarbonate fraction, NC_d).

As displayed in Fig. 63, the absolute cement content (Z_d) is expressed in a basic equation as the difference between original porosity (n_o), compaction (K), and the absolute amount of rock porosity (n_d, expressed as a percentage of the primary sediment volume):

$$Z_d[vol\%] = n_o - K - n_d \quad . \tag{14}$$

The absolute porosity (n_d) is related to the rock porosity (n) as follows:

$$n_d[vol\%] = n(1 - 0.01K) \quad . \tag{15}$$

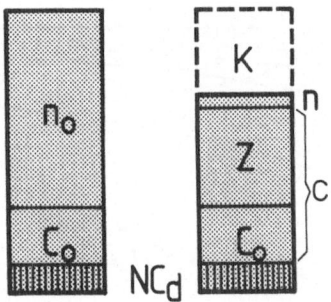

Fig. 63 Scheme for the calculation of the cement content (Z). Left: primary sediment with original pore space (n_o), original carbonate content (C_o), and normalized noncarbonate fraction (NC_d). Right: transformation into rock. The primary porosity (n_o) is greatly reduced by compaction (K) and precipitation of carbonate cement (Z). Due to this, the carbonate content of the rock (C) is enhanced as compared to the primary carbonate content (C_o).

If n_d in eq. 14 is substituted for by its equivalent expression in eq. 15, the following equation for the absolute cement content is obtained:

$$Z_d[vol\%] = n_o-n+K(0.01n-1) \quad . \tag{16}$$

Moreover, the absolute cement content can be expressed in terms of primary porosity (n_o), absolute clay content (NC_d), and carbonate content (C) if compaction (K) in eq. 16 is replaced by eq. 4:

$$Z_d[vol\%] = n_o-100 + \frac{NC_d}{1-0.01C} \quad . \tag{17}$$

In order to express the cement content as a percentage of the existing carbonate fraction, the carbonate volume (which is $100-K-n_d-NC_d$ of the primary sediment volume) must correspond to 100%. Therefore, the relative cement content (Z_c) is written as follows:

$$Z_c[vol\%] = \frac{100Z_d}{100-K-n_d-NC_d} \quad . \tag{18}$$

The terms n_d and Z_d have to be substituted for by eqs. 15 and 16,

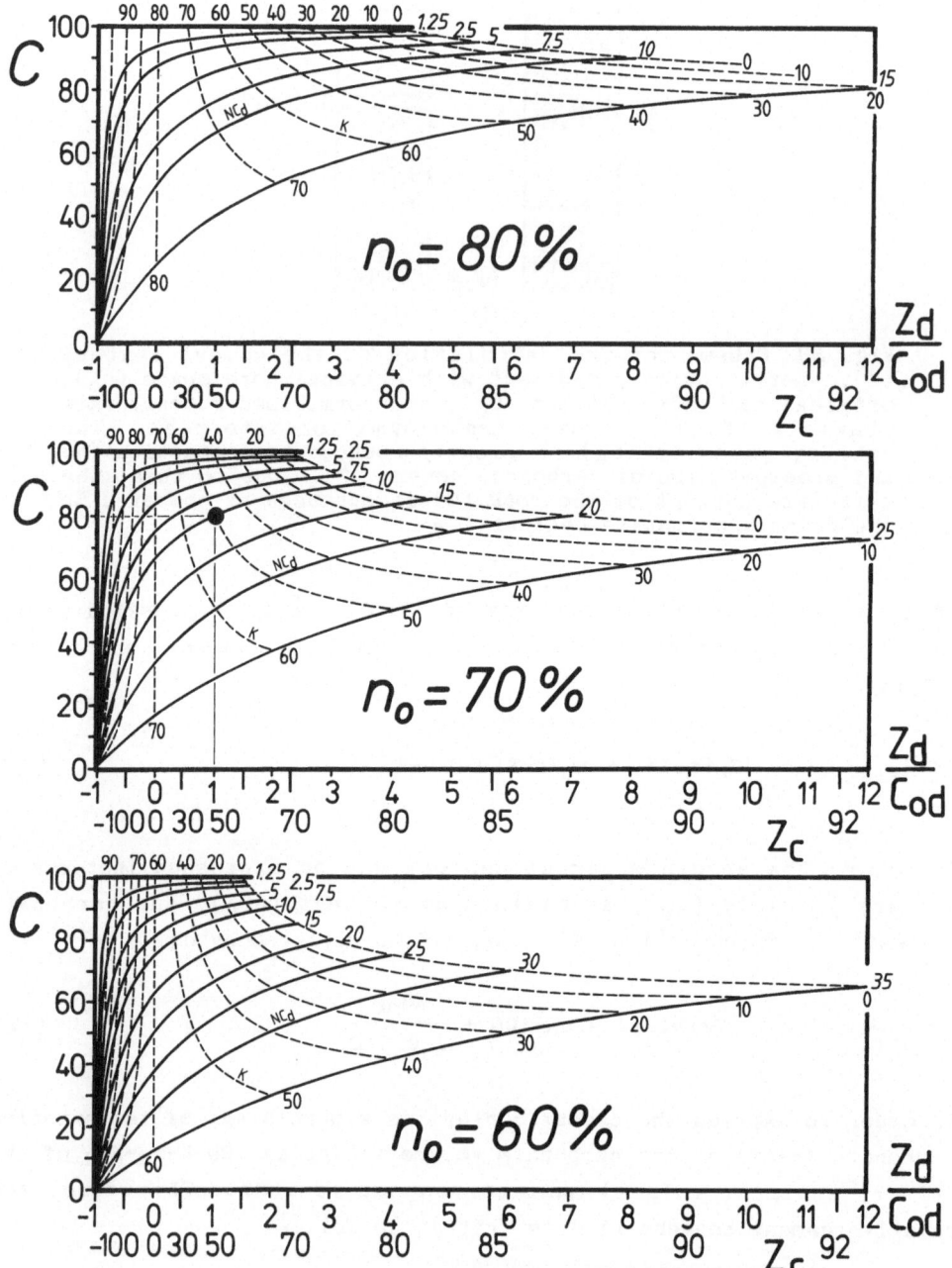

Fig. 64 The diagram shows the ratio between the absolute values of cement and primary carbonate (Z_d/C_{od}) and the relative cement content (Z_c). The two values are calculated for various contents of post-diagenetic carbonate (C), of standardized noncarbonate (NC_d, ——), of compaction (K, ---), and of primary porosity (n_o). It is assumed that the porosity of the existing rock is small enough to be negligible.

respectively; the relative cement content (Z_c) is then:

$$Z_c[\text{vol\%}] = \frac{100n_o - 100n + K(n-100)}{100 - NC_d - n + K(0.01n - 1)} \quad . \qquad (19)$$

However, during application the determination of compaction is commonly more difficult than the determination of the carbonate content. For that reason, compaction in eq. 19 can be replaced by eq. 4 to give the relative cement content in relation to the primary porosity (n_o), absolute clay content (NC_d), and the existing carbonate content (C):

$$Z_c[\text{vol\%}] = \frac{NC_d + n_o - 100 + C(1 - 0.01n_o)}{NC_d C} \times 10^4 \quad . \qquad (20)$$

The relative cement content (Z_c) is related to the cement number (z):

$$z = \frac{Z_d}{C_{od}} = \frac{Z_c}{100 - Z_c} \quad , \qquad (21)$$

and

$$Z_c[\text{vol\%}] = \frac{100z}{1+z} \quad , \qquad (22)$$

where $z > (-1)$. For a quick estimation, both the relative cement content (Z_c) and the cement number (z) are graphically represented in Fig. 64. The curves are applicable only to dense rocks, because the existing (post-diagenetic) porosity was assumed to be negligible.

In addition to the cement content, the primary carbonate content of the sediment (C_o) can be determined:

$$C_o[\text{vol\%}] = \frac{C_{od}}{1 - 0.01n_o} \quad , \qquad (23)$$

◁ Example: A given sample of lithified rock with a carbonate content of 80% contains as much cement as primary carbonate ($Z_d/C_{od}=1$) or has a relative cement content of $Z_c=50\%$ of the total carbonate. This is, if the primary porosity is $n_o=70\%$, and when the standardized clay content (NC_d) or the degree of compaction (K) is 10 and 50%, respectively.

where the normalized primary carbonate content (C_{od}) is derived from eq. 21:

$$C_{od}[vol\%] = \frac{100Z_d}{Z_c} - Z_d = 100 - NC_d - n_o \quad . \qquad (24)$$

By replacing C_{od} in eq. 23 by eq. 24, the original carbonate content (C_o) is dependent only on the absolute clay content (NC_d) and the original porosity (n_o):

$$C_o[vol\%] = \frac{100 - NC_d - n_o}{1 - 0.01n_o} \quad . \qquad (25)$$

Additional methods concerning the original carbonate content are presented in section 6.6.

5.4 Simulation of Diagenetic Separation

The simulation of the diagenetic separation of the original sediment into carbonate-rich and clay-rich layers is based on the carbonate compaction law and on the cement formulas. In principle, the compaction law (eq. 4) can be represented as a three-dimensional diagram comparing carbonate content, porosity, and compaction in a given sediment or rock sample with a constant absolute clay content (Fig. 65). By doing this, the curved relationship between carbonate content and compaction (which is valid for limestones with low porosities, see Fig. 7) changes to a spheroidal segment representing all possible stages of diagenesis to which a given sediment or rock sample may be subjected. Therefore, different types of sediment-rock transformations can be described with the carbonate compaction law. On the diagram, sediments are located at the lower edge, since they have high porosities, low degrees of compaction, and different carbonate contents. Rocks, however, are located at the upper edge of the spheroidal diagram since they display low porosities and both high and low amounts of compaction and carbonate (Fig. 65).

Carbonate diagenesis follows three basic processes: mechanical compaction, cementation, and chemical compaction (SCHOLLE, 1977; BATHURST, 1980a,b; HARRIS et al., 1985). All of these can be simulated using the three-dimensional diagram of the carbonate compaction law.

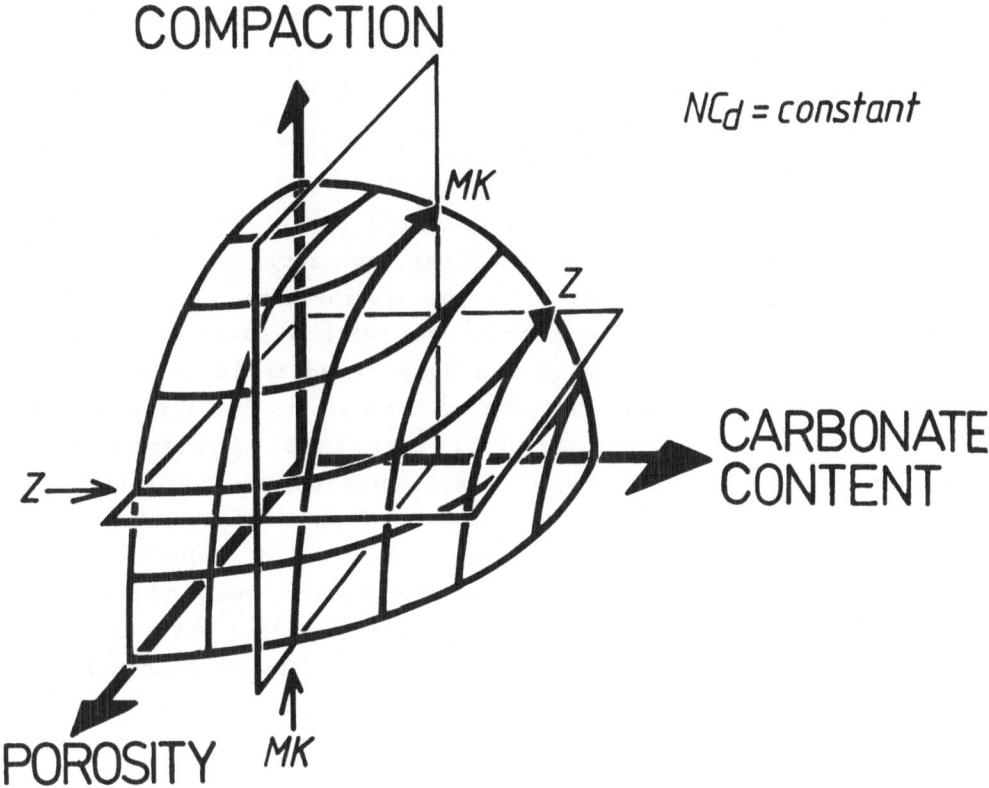

Fig. 65 Three-dimensional diagram of the compaction law. Mathematical relationship between all possible values of carbonate content, compaction, and porosity for a given sediment or rock sample with a constant absolute noncarbonate fraction (NC_d) during its diagenetic history. Mechanical compaction (MK) follows a curve parallel to the plane between the porosity and compaction axis, while cementation (Z) follows a curve parallel to the plane between the carbonate content and porosity axis.

Mechanical Compaction (MK): During mechanical compaction, only porosity is reduced, while the carbonate content remains constant. Curves of mechanical compaction (e.g., Fig. 65 curve MK) are parallel to a plane between the n and K axes of the diagram.

Cementation (Z): Compaction usually remains constant due to the strengthening and welding of grain contacts; porosity decreases, whereas carbonate content increases. Therefore, curves simulating the cementation process (e.g., Fig. 65 curve Z) are parallel to a plane between the n and C axes.

Chemical Compaction (-Z): Chemical compaction, or pressure dissolution, causes both a diminution of the $CaCO_3$ content and a further decrease in porosity. Thus, on the spheroidal diagram of the compaction law, chemical compaction results in left-trending curves (for example, see Fig. 66 a curve -Z).

In the following discussion, these three basic diagenetic curves are applied to simulate carbonate content, porosity, and compaction in the diagenetic history of marl-limestone alternations. As the example in Fig. 66a shows, diagenesis initiates with a phase of mechanical compaction; therefore, porosity and compaction follow the curve of mechanical compaction (MK, Fig. 66a). Thereafter, lithification sets in producing alternating zones of carbonate cementation and dissolution. Hence, carbonate content, porosity, and compaction must change along the branches of the two curves Z and -Z (Fig. 66a). They represent the sites of maximum cementation and dissolution (in the middle of what are later limestone and marl layers, respectively). In this model, all other carbonate and compaction values which oscillate in an alternation lie, for simplification, on lines with constant porosity connecting the branches Z and -Z (this is represented by outlined arrows or "tie lines" for two diagenetic stages in Fig. 66a). When the process of lithification ends, and porosity is extremely low; the relationship between carbonate content and compaction is similar to that obtained from measurements in marl-limestone alternations (see the summarizing graphs in Table 12).

The paths of the MK and Z curves (representing mechanical compaction and cementation, respectively) are easily calculated using the compaction law with a constant carbonate content and a constant degree of compaction, respectively (section 2.2). However, the determination of curve -Z (chemical compaction) is based on the fact that the dissolution zones are the only source of the carbonate which is reprecipitated in the pore space of the cementation zones or limestone layers (convincing evidence was found that diagenetic bedding occurs predominantly in a closed carbonate system, section 2.3.1). This implies that in every diagenetic stage with a given

Fig. 66 Formation of diagenetic carbonate oscillations explained by using the carbonate compaction law with normalized noncarbonate fractions of $NC_d=5\%$ (a) and 30% (b). MK=mechanical compaction, Z=cementation, -Z=chemical compaction or pressure dissolution, 0 to 4=successive diagenetic stages. The two blank tie lines in Fig. 66a represent fluctuations in carbonate content and compaction for an early and a late diagenetic stage.

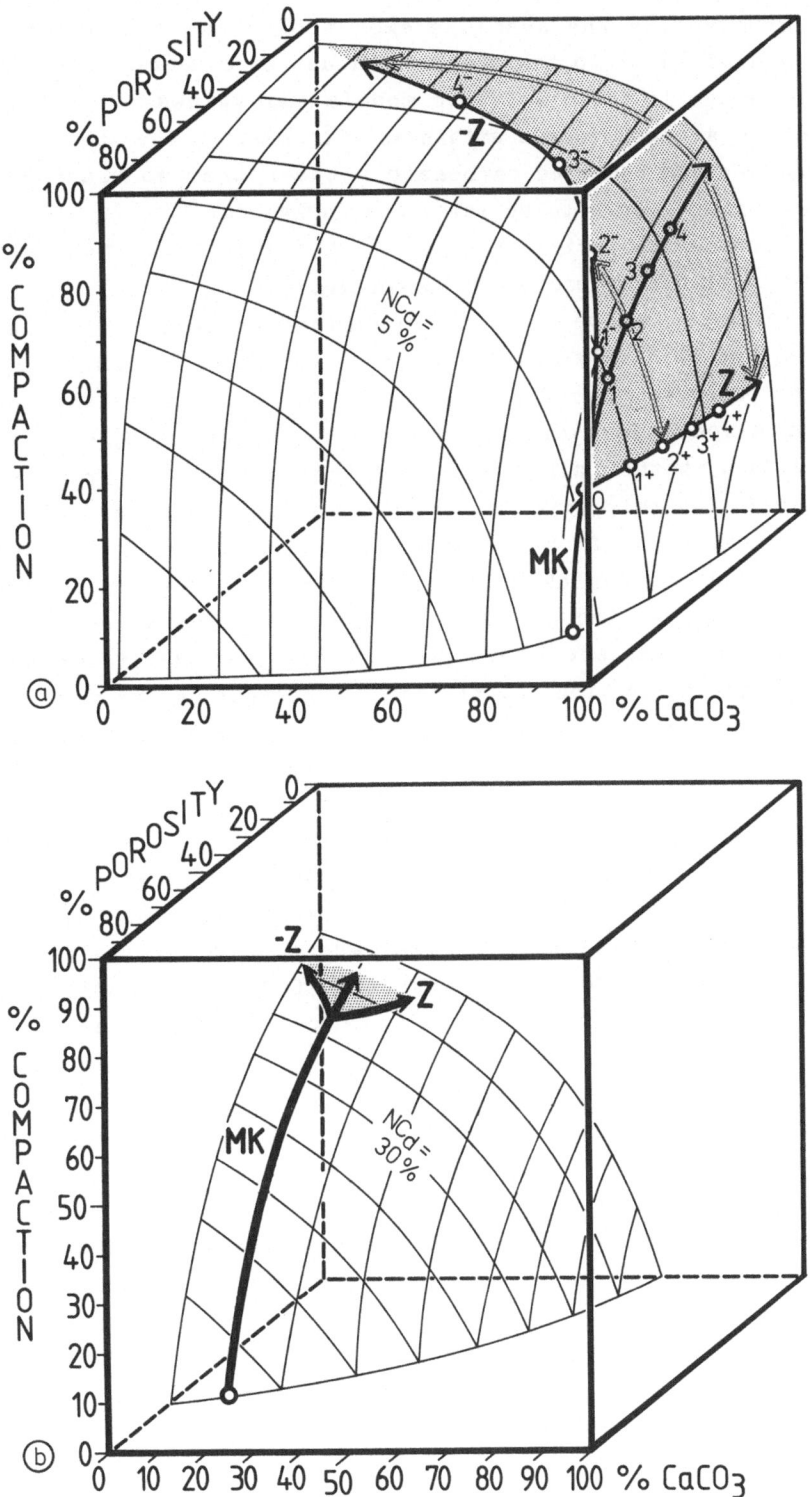

constant porosity the absolute amount of dissolved carbonate ($-Z_d$) must be equivalent to the absolute amount of cemented carbonate (Z_d). However, since in the simulation models all values of carbonate and compaction are considered within an alternation, the absolute amounts of dissolved and cemented carbonate between the two branches of the curves Z and -Z have to be integrated using eq. 17. In addition, one has to consider the original ratio of the layer thickness which are relatively richer or poorer in carbonate.

Diagenetic bedding in fine-grained carbonate-rich sediments (NC_d=5%) can be easily simulated, since the model predominantly agrees with the observations (Fig. 66a). On the other hand, one can also develop a model for very carbonate-poor sediments (NC_d=30%) in order to ascertain whether diagenetic bedding can also occur there (Fig. 66b). Unfortunately, the latter model is somewhat speculative because it is interpolated from the studied, carbonate-rich alternations. Diagenesis of clay-rich sediments presumably follows a long curve of mechanical compaction parallel to the left edge of the spheroidal diagram of the compaction law (see the onset of lithification in Fig. 60). Theoretically, the relative cement content can be very high, if cementation began early (e.g., during the formation of early diagenetic carbonate concretions). However, in the case of early cementation, the necessary amount of cement can not be provided directly by dissolution in the adjacent marl layers, because the carbonate content is only low. The diagenetic carbonate system would be either open, or, if closed, cemented layers in marly alternations with early diagenesis should have large shale intervals. However, because carbonate-poor sediments (with $CaCO_3$ below 30%) presumably undergo intense mechanical compaction before carbonate redistribution sets in little pore space is left to be cemented. Thus, in primary clay-rich sediment, carbonate redistribution must usually be low and should likewise occur in a closed system. Contrary to carbonate-rich sequences, which are heavily affected by the processes of diagenetic bedding, the primary bedding rhythm of carbonate-poor alternations should be essentially preserved. For instance, SCHNEIDER (1964) described such bedding cycles in the Lower Cretaceous shales of northern Germany.

In the following pages, the formation of diagenetic bedding, which occurs in fine-grained sediments with different carbonate contents, is represented by three models (Figs. 67a,b and 68). They are based on the following assumptions:

1. Primary porosities (only fine-grained sediments are included) should vary between 65% (at 100% $CaCO_3$) and 80% (at 0% $CaCO_3$, see section 2.3.2).

2. The onset of lithification should follow the regression curve portrayed in Fig. 60a.

3. The existing dissolution zones should originally have been thicker than the precursor of the present cementation zones. In the original sediment column, the ratio of beds relatively richer to those relatively poorer in carbonate is assumed to be 1:2.

4. Porosity differences between marl and limestone layers are ignored. For reasons of simplicity, it is assumed that at every diagenetic stage (during the progressively reduction of pore space) there is a constant porosity in the developing marl and limestone layers (which actually is not true, see section 5.2).

Although the three models portrayed in Figs. 67a,b and 68 are based on a somewhat arbitrary data input and on extrapolation from data on carbonate-rich sediments, the diagrams give important insights into the development of diagenetic bedding. The first model (Fig. 67a) shows diagenetic separation in sediments with different absolute clay contents (NC_d=1 to 15%), which is represented in a similar manner as those described in Fig. 66. Diagenetic separation into layers of cementation (curve Z) and chemical compaction (curve -Z) occurs after a phase of mechanical compaction (curve MK). Successive diagenetic stages in Fig. 67a are simply expressed by tie-lines which connect the curves Z and -Z. The second model, Fig. 67b, is compiled from Fig. 67a. It displays carbonate content, compaction, and porosity for three zones of diagenesis: the predominantly two dimensional zone of mechanical compaction (MK) and the two areas of carbonate cementation (Z) and carbonate dissolution (-Z). In model three (Fig. 68), the more abstract diagrams in Fig. 67a,b are transformed into a graph which schematically displays the development of diagenetic carbonate curves for several diagenetic stages. For reason of simplicity, no primary (slight) carbonate oscillations are shown in the three diagrams.

Results:

1. Carbonate-rich primary sediments are subjected to only a short phase of mechanical compaction and usually undergo an early onset

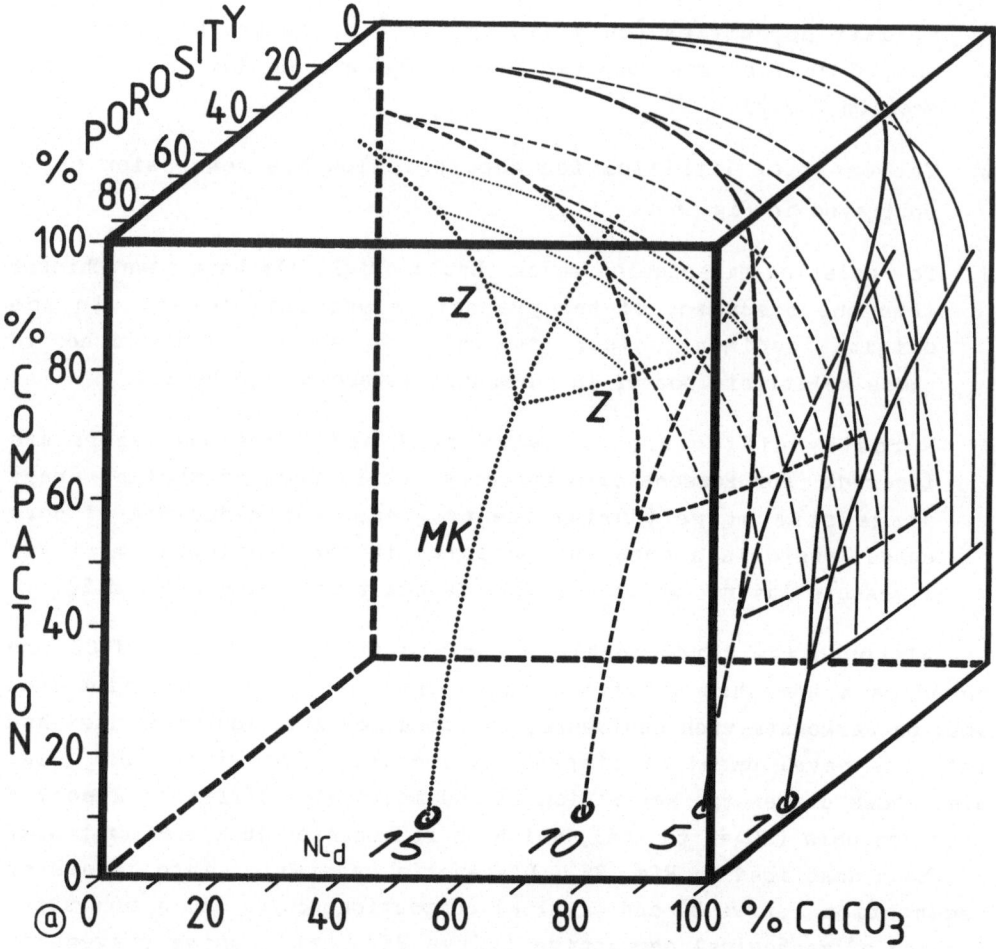

Fig. 67 a: Model for the production of diagenetic bedding
in fine-grained, calcareous sediment based on the carbonate
compaction law with standardized noncarbonate fractions of
NC_d=1 to 15%.   Shown are the main diagenetic curves
(MK=mechanical compaction, Z=cementation, -Z=chemical
compaction or pressure dissolution).   Thin tie lines between
the Z and -Z curves represent the oscillations of carbonate
content and compaction in successive stages of diagenesis.
b: Block diagram for fine-grained, calcareous sediments as
presented in Fig. 67a.   Striped plane is the zone of
mechanical compaction (MK), and, after the onset of
lithification, the striped plane is the neutral zone between
the areas of cementation (Z) and chemical compaction (-Z).
Small numbers indicate various successive diagenetic stages
(0 to 5).

of diagenetic carbonate redistribution.   This enables the
precipitation of large amounts of cement in the limestone layers,
triggers intense carbonate dissolution in the marl beds, and
greatly enhances the primary bedding rhythm (see sections 4.4.3

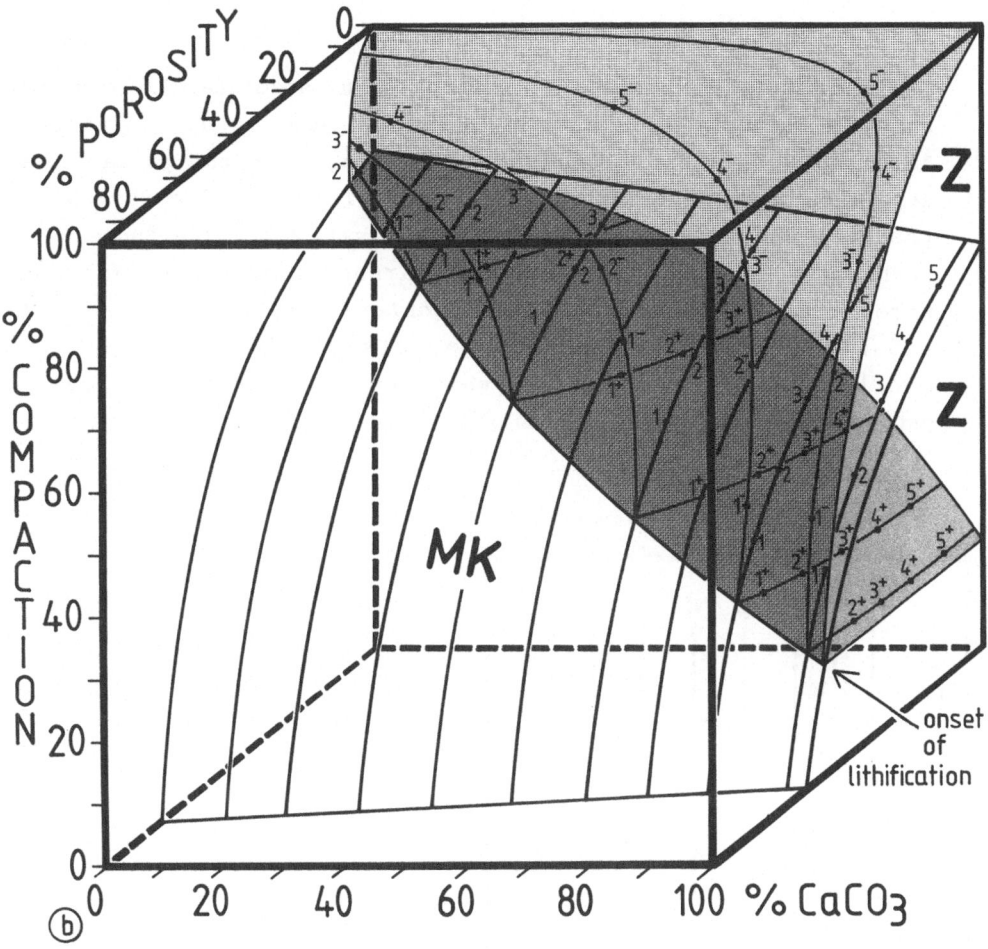

and 4.6). However, after the onset of lithification, carbonate redistribution predominantly affects compaction but does not significantly alter the carbonate content. This is due to the fact that the relative carbonate content (expressed in percent of the total amount of solids) cannot be changed in any substantial amount by an increase or decrease of the absolute amount of carbonate, provided that the absolute clay content is very low. Hence, in spite of an early onset of lithification and high absolute cement values (see Table 12), diagenetic carbonate oscillations form only at late stages of diagenesis after a considerable number of dissolution-reprecipitation episodes, when the pore space is already very low (Figs. 67a, 68). Diagenetic differences in CaCO$_3$ are predominantly caused by carbonate dissolution in the thin, stylolitic marl beds, whereas cementation enhances the high primary carbonate content only very slightly in

the limestone layers.

2. Sediments with intermediate carbonate and absolute clay contents have a relatively long phase of mechanical compaction and often a late beginning of diagenetic separation, when the primary pore space is already substantially reduced (Fig. 67). However, since the absolute clay content is comparatively high, the increase or decrease in the absolute amount of carbonate does significantly affect the (relative) carbonate content. Although the absolute amount of cement is low, diagenesis results in large $CaCO_3$ oscillations (see Table 12), which begin to generate simultaneously with the onset of lithification.

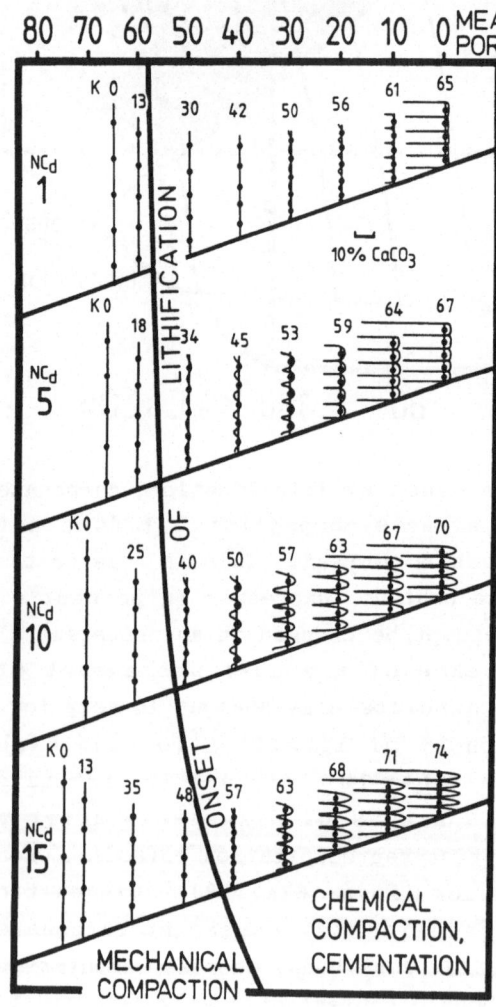

Fig. 68 Formation of diagenetic carbonate oscillations in fine-grained, calcareous sediments with standardized noncarbonate fractions of $NC_d=1$ to 15%, which is presented in relation to the mean porosity and the amount of compaction (K). Shown are schematic carbonate curves, which are representations of how diagenetic carbonate curves successively fluctuate around centers of lithification (●). Data is based on the models in Fig. 67.

3. As already discussed, the very clay-rich sediments, which this
study does not deal with, should be little affected by diagenetic
bedding because of high reduction in porosity due to mechanical
compaction (Fig. 66b).

In spite of the early beginning of diagenetic carbonate
redistribution in some sediments, significant carbonate oscillations
are only formed during late stages of diagenesis when the pore space
is considerably reduced (below porosities of 30 to 40%); regardless of
whether or not the primary sediment was carbonate-rich. This is
supported by data obtained from the Deep Sea Drilling Project. In
pelagic carbonates, the formation of bedding-parallel stylolites
begins when porosity is between 5 and 40% and overburden ranges from
600 to 1000m (Table 16).

Table 16: Onset of the development of bedding-parallel stylolites in
pelagic carbonate sequences (DSDP).

Site	Meters of overburden	Carbonate content (%)	Porosity (%)	Reference
367	1050	60-90	5-20	GARDNER et al., 1977
463	632	90	20-40	THIEDE et al., 1981
516	875	75-90	30	BARKER et al., 1984

The effect of diagenetic bedding on sediments with different
carbonate content is summarized in Fig. 69. The clay-rich bedding
types are extrapolated from the studied alternations containing medium
to high carbonate contents. Only the carbonate-rich alternations
weather to distinct marl and limestone layers, whereas the $CaCO_3$
alternations in carbonate-poor rocks weather completely to marl,
because their carbonate content is below that of the marl-limestone
weathering boundary (C_W). In all cases, the total compaction is
relatively high. This is because carbonate is redistributed
predominantly within a closed system; therefore, the amount of total
compaction in lithified alternations must be on the same order as the
amount of original pore volume. Hence, in spite of large compactional
differences in alternations with diagenetic bedding, their total
amount of compaction should be essentially the same as that observed
in neighboring shales.

Rhythmicity, differential compaction, the constancy of the maximum
carbonate content, and the ability to develop angular-shaped maximums
on the carbonate curves increase with increasing primary carbonate
content. However, the largest diagenetic carbonate oscillations are

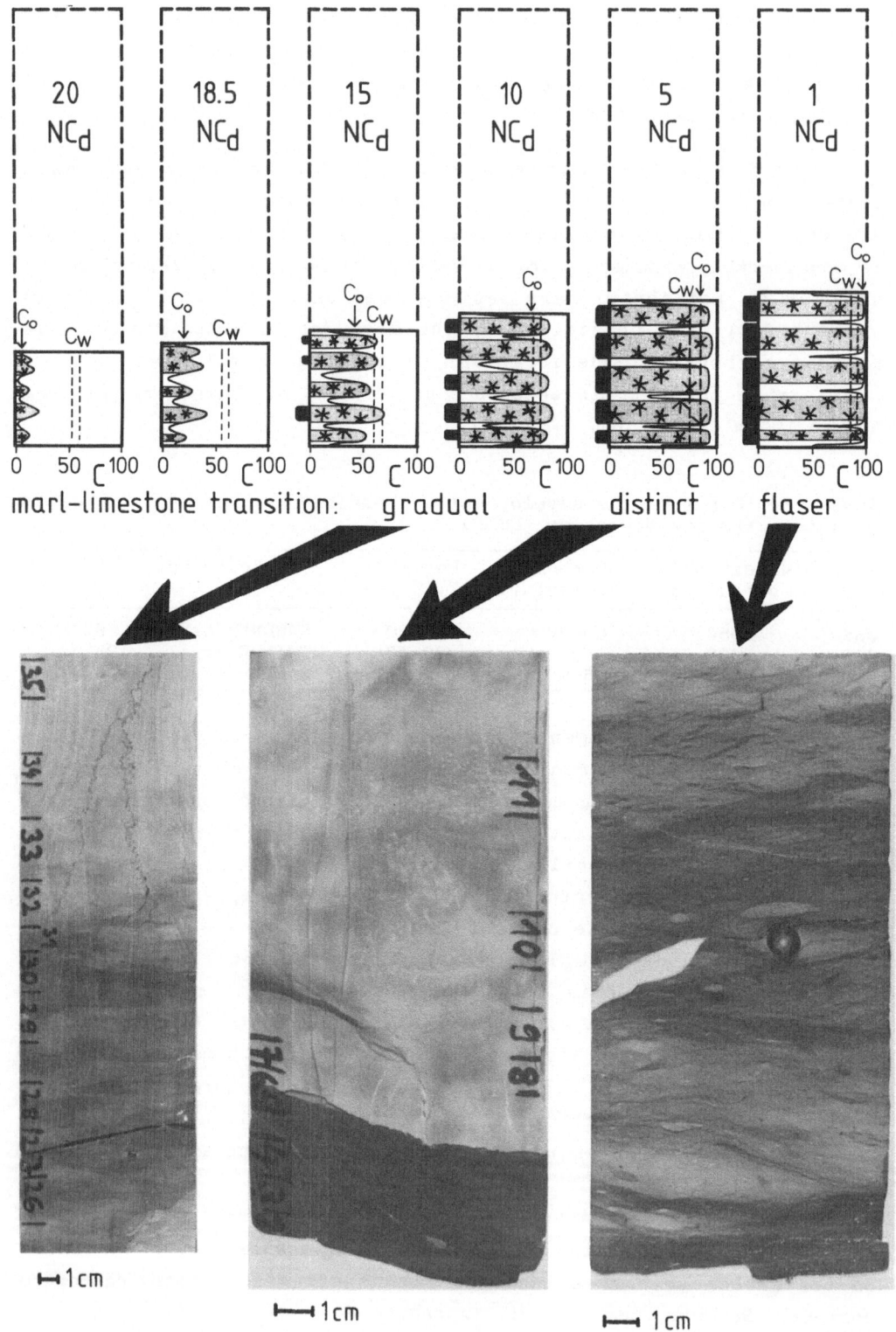

20 NC$_d$	18.5 NC$_d$	15 NC$_d$	10 NC$_d$	5 NC$_d$	1 NC$_d$

marl-limestone transition: gradual   distinct   flaser

⊢1cm   ⊢1cm   ⊢ 1cm

found in alternations with intermediate carbonate content.

The position of the neutral carbonate value (C_n) on the different types of the carbonate curves greatly influences the appearance of the transition between dissolution and cementation zones (Fig. 69, lower part). As mentioned above, the shapes of the carbonate maxima change with increasing mean carbonate content from sinusoidal to angular. Moreover, the neutral carbonate value between dissolution and cementation zones increases with increasing carbonate content (see Fig. 43). Therefore, in clay-rich alternations, the neutral value is located where the sinusoidal carbonate curve decreases rapidly towards the middle of the marl bed. For that reason, the transition occurs within the (weathered) marl bed at a relatively low $CaCO_3$ content; hence, in marl-rich alternations the macroscopic transition between dissolution and cementation zones is gradual. However, in highly calcareous alternations with angular carbonate curves, the neutral carbonate value is still inside the (weathered) limestone ledge, because the neutral value occurs at a high $CaCO_3$ content, which is the same order of magnitude or even greater than the carbonate content of the weathering boundary. Therefore, in carbonate-rich alternations the transition between dissolution and cementation zones displays flasery, stylolitic seams (these are equivalent to the "sutured- and nonsutured-seam solution" types described by WANLESS, 1979). The transition between both zones consists of a more or less distinct contact when a sharp decrease in the carbonate curve corresponds to the neutral carbonate value. This is usually at medium to high mean contents of carbonate and absolute clay (Fig. 69).

The preservation of delicate calcareous fossils in the weathered marl beds depends on the carbonate content of the neutral zone and that of the weathering boundary. Since the weathered marl beds are usually thicker than the dissolution zones, the marl beds can contain delicate fossils which are unaffected by pressure dissolution. In the flasery marl joints of highly calcareous alternations, dissolution is

Fig. 69 A summary of the effects of diagenetic bedding for fine-grained, calcareous sediments with standardized noncarbonate fractions of NC_d=1 to 20%. Shown are the original volume of sediment (---) and the post-diagenetic compacted rock volume (——) with its typical carbonate curve (C) and the mean primary composition (C_0) of the original sediment. Also represented is the carbonate content at the weathering boundary (C_w, dashed double line), the weathering profile (black, left) and the extent of dissolution (blank) and the cementation zones (starred, shaded) in the sections. Marl-limestone transitions from KB6, Strassberg, southern Germany (Upper Oxfordian).

often intense. However, delicate fauna might be preserved in small areas of carbonate precipitation which occur around massive fossils or between certain dissolution seams.

5.5 Diagenetic Bedding: A Predominantly
Stratiform Process

Although low angles of disparity in dip between the original and diagenetic bedding may occur on a small scale (WALTHER, 1983; SIMPSON, 1985), diagenetic bedding is mainly parallel to the original bedding. This does not exclude, however, the fact that diagenetic bedding often results in selectively cemented individual beds which can be traced several 100's of km, in some cases. For instance, HATTIN (1971, 1985) and ELDER & KIRKLAND (1985) traced individual and partly concretionary limestone layers of the Upper Cretaceous Bridge Creek Limestone in the Western Interior of North America over 700km using isochronous bentonite beds as controls. Other evidence for a mainly stratiformal process of diagenetic bedding is that certain beds in the primary stratification are "copied" by diagenetic marl and limestone layers even when the primary beds are deposited at an angle to the direction of the major stress which was caused by overburden of sediment (BATHURST, 1984). This phenomenon can be observed in Neuffen (see Fig. 90), Gubbio, and Angles where crumbly slump layers and point-bar beds in submarine channel fills are "copied" by diagenetic bedding. The only location this author knows where a special type of secondary bedding plane cuts the primary stratification is at Stevens Klint near Copenhagen, Denmark. There, the primary biostromal bedding in the Maastrichtian chalk is cut at low angles by delicate, but traceable stylolitic joints, at regular intervals of about 40cm. The primary bedding can be inferred from layers containing "bands" of chert concretions. Perhaps, the secondary marl joints are due to an overburden of approximately 2000m of ice during the Pleistocene.

The tectonic phenomenon of dissolution cleavage in chalks and limestones shall be mentioned briefly (e.g., PLESSMANN, 1964; ALVAREZ et al., 1978; GEISER et al., 1981). Usually, wavy dissolution and shear planes develop at a high angle to the bedding. For the most part, the dissolved carbonate is not reprecipitated within the pore space in the laminae between the slate planes, but is transported over long distances (MIMRAN, 1977).

5.6 Discussion of the Seibold Model

In a classic investigation, SEIBOLD (1952) studied the rhythmic alternations of the Upper Jurassic in southern Germany and was the first to develop a quantitative model of deposition in marl-limestone alternations. This so-called "Seibold model" has become widely used (e.g., HILLER, 1964; FLÜGEL & FENNINGER, 1966; FLÜGEL, 1968; FÜCHTBAUER & MÜLLER, 1977, BAUSCH et al., 1982). However, according to the results of this study, the Seibold model should be considerably modified.

SEIBOLD (1952) assumed that marl-limestone alternations are not affected by diagenetic carbonate redistribution processes or differential compaction. Further, he assumed that the same time is required for the deposition of the existing marl and limestone layers. Based on these assumptions, he evaluated the "absolute amount of clay per bed" which he expressed in cm thickness of a given marl or limestone layer, respectively. It must be stressed that SEIBOLD's "absolute clay content" is different from the normalized or absolute clay content (NC_d) used in this study. Here, the normalized clay content is expressed as a percentage of the original sediment volume. SEIBOLD's calculations demonstrated that in the Upper Oxfordian marl-limestone alternation the absolute clay content per bed thickness remained predominantly constant, but that the absolute carbonate content varied considerably (Fig. 70a). Since SEIBOLD assumed that diagenetic carbonate redistribution was negligible, he inferred that rhythmic carbonate deposition was superimposed on a constant rate of clay background sedimentation.

SEIBOLD's conclusion that the clay content per bed thickness in marl-limestone alternations is often constant must now be reinterpreted as a result of diagenetic carbonate redistribution (Fig. 70b). According to the compaction data from this study, marl-limestone alternations developed from an original sediment with a fairly uniform composition. Moreover, in the Upper Oxfordian of southern Germany, layers originally slightly richer and poorer in carbonate content had nearly the same thickness in the primary sediment (see Fig. 35). Hence, after differential compaction and, therefore, carbonate redistribution, the absolute clay content per bed must remain constant in each layer, while the absolute carbonate content must show pronounced variations. Since the studied lithified alternations have large compactional differences, the mainly constant

SEIBOLD MODEL

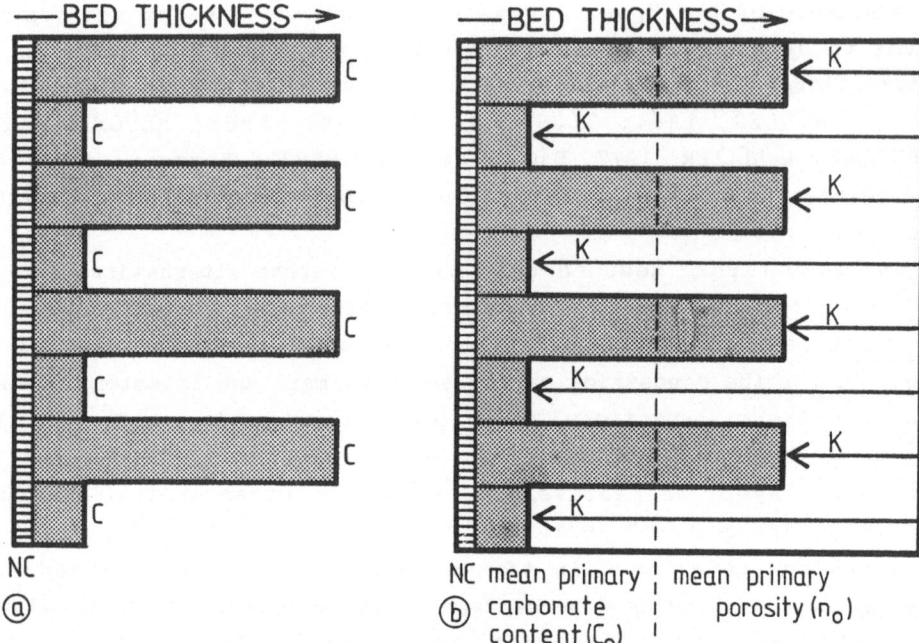

Fig. 70 a: Seibold Model for marl-limestone alternations.
In many alternations, the total amount of noncarbonate per
bed (NC, striped) is predominantly constant while the
carbonate content (C, shaded) varies considerably. If no
diagenetic redistribution of carbonate occurred, this
phenomenon is due to alternating periods with and without
$CaCO_3$ dilution of a constant clay background sedimentation.
The dilution would be caused by the rapid deposition of
calcareous sediment.
b: Interpretation in terms of diagenetic carbonate
redistribution. Layers, which were originally slightly
richer and poor in carbonate content underwent differential
compaction. In the slightly compacted layers, the remaining
pore space was filled with carbonate cement which was
provided by the highly compacted marl beds as a result of
pressure dissolution. Assumed mean primary sediment
composition shows the amounts of noncarbonate (NC), original
carbonate (C_0), and original porosity (n_0). Note, that the
noncarbonate content per bed remains unchanged during the
diagenetic redistribution of carbonate.

absolute clay content per bed found by several authors in
marl-limestone alternations now confirms the concept of diagenetic
bedding rather than the assumed high carbonate oscillations in the
primary sediment. In the future, the Seibold model should no longer
be implemented, because its sedimentological validity is severely

restricted. If one expresses the clay content relative to the thickness or volume of the marl and limestone layers, several problems arise:

1. The absolute clay content per bed is not clearly related to primary depositional parameters, because the existing marl and limestone beds underwent different degrees of compaction. On the other hand, the clay content normalized to the primary sediment volume used in this study is a valid parameter which can be clearly interpreted in terms of deposition and diagenesis.

2. The thickness of the weathered marl and limestone layers corresponds to the thickness of the dissolution and cementation zones only in highly calcareous alternations (see Fig. 43). In marl-rich alternations, the marl layers are usually somewhat thicker than the dissolution zones; additionally sites containing small carbonate maxima can weather totally to marl (see Fig. 32).

3. In the original sediment the precursors of the present dissolution and cementation zones did not have equivalent thicknesses. The thickness of what are the later cementation zones was between 34 and 50% of the primary sediment column.

The objections raised in problems 2 and 3 indicate that the absolute clay content (per bed thickness) increases in the marl beds of carbonate-poor alternations relative to that contained in the limestone layers. This effect is clearly visible in the marl-rich portions of the alternations studied by SEIBOLD (1952).

5.7 Conclusions

As the existence of calcareous pressure shadow structures show, the stratiformal dissolution-reprecipitation process of $CaCO_3$ is triggered by differential stress in the grain framework of beds slightly richer and poorer in carbonate content. The diagenetic transport and exchange of carbonate between adjacent marl and limestone beds is predominantly governed by diffusion. Appreciable cementation sets in once an overburden between 50 and 400m has been deposited and when original porosities have been reduced to values between 50 and 60%, depending on the carbonate content of the primary sediment. However, it is evident from simulation models that significant diagenetic

changes in the relative $CaCO_3$ content form only in late stages of
diagenesis when the porosity becomes less than 30%, especially in
carbonate-rich and carbonate-poor alternations. In the limestone
layers, cementation considerably reduces the pore space to 0 — 5%,
whereas the porosity in the marl layers remains relatively high (5 to
15%).

6 APPLICATIONS AND RAPID METHODS FOR THE QUANTIFICATION OF DIAGENETIC CARBONATE REDISTRIBUTION

The applications of the methods developed in this study include not only an analysis of the diagenesis in marl-limestone alternations, but of other calcareous sediments and rocks as well. This can be accomplished since the carbonate compaction law is a fundamental natural relationship which enables the diagenesis of common carbonates to be understood in terms of mass exchanges. For instance, if one can evaluate the absolute clay content of a sequence, compaction can be easily calculated from the readily determinable parameters: carbonate content and porosity. This measurement of compaction can then be applied to a sequence of calcareous rock to determine its cement content. Thus, in the future, compaction and cement logs taken from field sections and drill holes could become valuable, new tools in oil and gas exploration. Moreover, the chemical and isotopic composition of existing carbonates should be interpreted as a result of the diagenetic mass exchange, which is closely related to compaction. This new, more inclusive view of compaction should also be applied to other fields in geology, for instance in basin analyses and subsidence calculations. Although frequently ignored, compaction is one of the most important geological and diagenetic parameters. However, the application of precise methods (as presented in sections 2 and 5) is often lengthy and tedious. Therefore, rapid and somewhat imprecise methods are developed in the following section which rely on only a comparatively small number of samples.

6.1 Standardized Noncarbonate Fraction (NC_d)

The standardized noncarbonate fraction, which is expressed as a percentage of the original sediment volume (see section 2.2), is the basic parameter for the application of both the compaction law and the cement formulas. If no suitable structures are available to determine the amount of compaction and to calculate the standardized non-carbonate fraction (or the absolute clay content) by solving eq. 2, an estimation of the absolute clay content is carried out by using the following methods.

The absolute clay content is equivalent to the relative clay content if $CaCO_3$ cementation occurred very early in the diagenetic history (for instance, in early diagenetic concretions), because the original volume is assumed to be only slightly reduced by mechanical compaction prior to the onset of cementation (see Fig. 7). However, the absolute clay content is always less than the relative clay content if the sediment underwent compaction. According to the compaction law, in an interval with up to 50% compaction, the influence of compaction on carbonate content (or relative clay content) is small if the absolute clay content is low (see Fig. 7). Therefore, in carbonate-rich alternations (Type III), the noncarbonate fraction in the middle of the limestone layers (that is, the site containing the lowest amount of compaction) gives an approximation of the absolute clay content in the sequence.

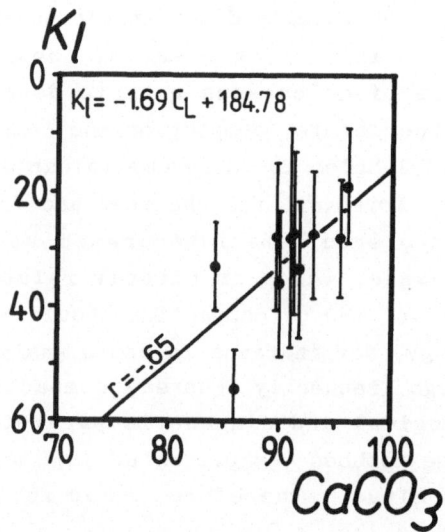

Fig. 71 Relationship between compaction at the onset of lithification (K_1) and the mean post-diagenetic carbonate content in the middle of the limestone layers.

In order to determine the absolute clay content in more marly alternations, the compaction at the onset of lithification (K_1) must be ascertained by using the carbonate content in the middle of the limestone layers (C_L, Fig. 71):

$$K_1 = -1.69C_L + 184.78 \quad . \tag{26}$$

The author suggests that one should first determine the carbonate content in the middle of five to ten limestone layers in an alternation and then estimate the degree of mechanical compaction using eq. 26. Then, the absolute clay content can be determined using eq. 2 or 3. Eq. 26 cannot be applied to semilithified alternations with a porosity of more than 30%.

6.2 Porosity in Lithified Alternations (n)

Errors which result from small differences in porosity are negligible in utilizing the rapid methods, so long as the total porosity is below 30% (see the compaction law, Fig. 7). If necessary, the porosity in lithified, micrictic alternations can be estimated because it increases regularly with decreasing carbonate content (see section 5.2). In the following expression, n is the measured porosity and C is the carbonate content (see Fig. 62b).

$$n = 5.56 \ln(100-C) - 9.46 \quad . \qquad (27)$$

6.3 Compaction (K)

If it is impossible to determine compaction directly or indirectly as previously discussed (see section 2.1.2), compaction can be calculated using eq. 4 and the curves in Fig. 7. The absolute clay content can then be estimated according to the methods presented in section 6.1, and rock porosity can be ascertained with eq. 27.

6.4 Primary Porosity (n_o)

An approximation of the primary porosity can be obtained by a quantitative decompaction of ten samples. According to the methods presented in sections 6.2 and 6.3, compaction and porosity are determined in the middle of five marl and five limestone layers which form an individual sequence. For every sample, an apparent primary porosity must be determined which does not represent the original porosity due to either dissolution or cementation of $CaCO_3$. The apparent primary porosity results from the addition of compaction to the normalized porosity (n_d, eq. 15). If diagenetic carbonate

redistribution occurred in a closed system, the mean value of the
apparent porosity in the marl and limestone samples should be
equivalent to the mean decompaction porosity in the studied section.
However, in the original sediment, the precursor of the existing marl
beds was on an average one and a half times thicker than the precursor
of the present limestone layers (see Table 11). Therefore, the
apparent porosity of the marl layer samples must be multiplied by a
factor of 1.5 before a mean primary porosity can be calculated. The
error of this method is about $\pm 10\%$ of the primary porosity when it is
determined using the more detailed decompaction calculation in section
2.3.5.

6.5 Cement Content (Z_c, Z_d)

The relative and absolute cement content (Z_c and Z_d, respectively) can
be estimated by using the diagrams in Fig. 64, or by calculating eqs.
16 and 17, or eqs. 19 and 20. The primary porosity (section 6.4) and
two other parameters must be known in order to perform the
calculations: such as carbonate content, compaction (section 6.3),
the standardized noncarbonate fraction (section 6.1), and the existing
porosity (section 6.2).

6.6 Primary Carbonate Content (C_o)

In principle, it is possible to ascertain the average carbonate amount
contained in the primary sediment of marl-limestone alternations by
performing a few carbonate determinations. They have to be carried
out at the weathering boundary, because in some cases the carbonate
content at the weathering boundary corresponds to that at the neutral
boundary between the dissolution and cementation zones. However, a
considerable error results from using this method, if the carbonate
content at the weathering boundary and that at the neutral zone are
not quite equal (see Fig. 43). Moreover, the most pronounced $CaCO_3$
change within the rock column occurs in the neutral zone (see Figs.
46, 69). For example, in carbonate-rich alternations (Type III), two
sampling sites only one centimeter apart can have carbonate
differences of up to 40%! Therefore, the direct estimation of the
mean primary composition by using the carbonate content at the
weathering boundary should only be performed in Type I, marl-rich

alternations, which display sinusoidal carbonate oscillations (section 4.1). In all other cases, an indirect determination method is proposed. This can be accomplished by applying eq. 25, which uses absolute clay content (section 6.1) and primary porosity (section 6.4).

A graphic method for the estimation of the original $CaCO_3$ content in fine-grained carbonates is derived in Fig. 72a. In this case, the process of diagenetic separation of a primary sediment into carbonate-rich and clay-rich beds is expressed in a similar manner as in Fig. 66a, but here the curves are projected onto the plane which stretches between the axes of carbonate content and compaction. Additionally, Fig. 72a is rotated 180° relative to its position in Fig. 66a, hence the degree of compaction increases from top to bottom of the diagram. The diagenesis of marl-limestone alternations first involves mechanical compaction and then carbonate redistribution. The beginning of the carbonate redistribution process is defined by a curve (given in Fig. 60a) describing the amount of mechanical compaction (K_1) in relation to the primary carbonate content. At a later stage of diagenesis, when most of the pore space is closed, all values of carbonate and compaction lie on a curve which is equivalent to the compaction law for low porosities (see Fig. 7). However, the entire length of the theoretical carbonate compaction curve is not represented by the measured data (see Table 12). It is evident that the vertical part of the carbonate compaction curve in Fig. 72a is more complete if mechanical compaction is low or the primary carbonate content is high.

Fig. 72b is an extension of Fig. 72a. It shows how the different compositions of the primary sediment (with varying amounts of absolute clay but with a constant porosity of 80%) determine the post-diagenetic amounts of carbonate and compaction on the vertical position of the theoretical curve described by the compaction law. The length of this curve which is represented by the measured data is directly related to the amount of mechanical compaction, which depends in turn on the primary sediment composition. This relationship enables one to use a graphic estimation of the primary carbonate content.

If one knows the amounts of carbonate and compaction in any given fine-grained sample, one can find the corresponding carbonate compaction curve given for a certain absolute clay content in Fig. 72b; then one follows this curve until one reaches the specific amount of mechanical compaction (arrows). Thereafter, one proceeds to the

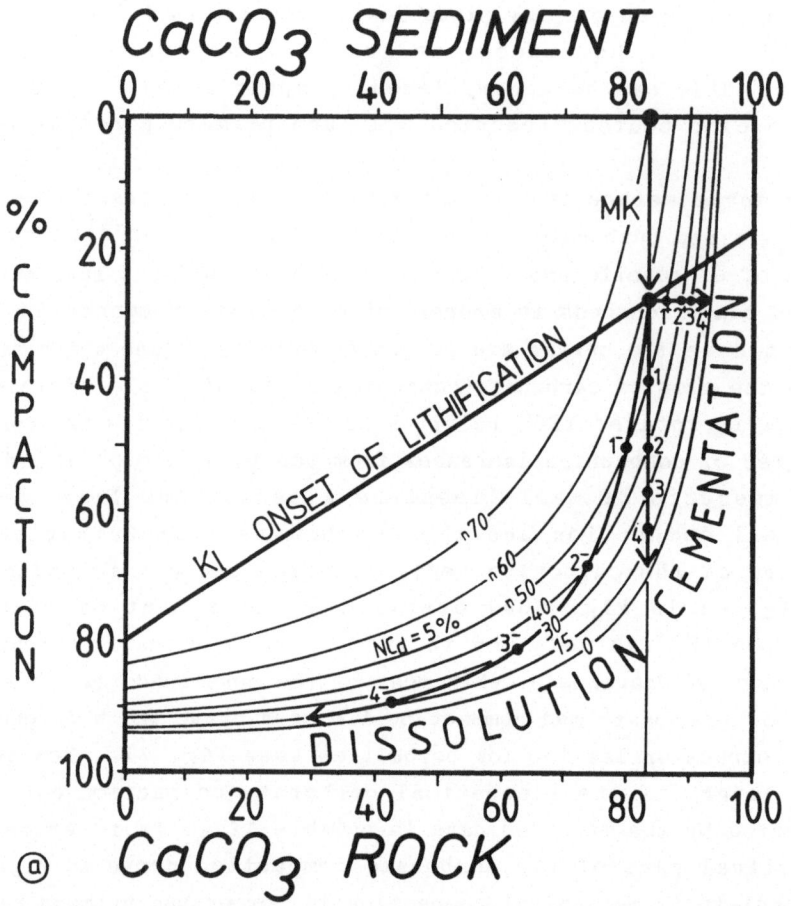

Fig. 72 a: Origin of diagenetic carbonate oscillations similar to that presented in Fig. 66. Projection of curves is onto the plane between the axes of carbonate content and compaction. MK=mechanical compaction, K_1=compaction at the onset of lithification as calculated in Fig. 60a, curves: theoretical relationship between carbonate content and compaction after the compaction law calculated for various porosities (n=0 to 70%).
Example: Sediment with 83% $CaCO_3$, a primary porosity of 70%, and with a normalized noncarbonate content of NC_d=5%.
b: Diagram for the estimation of the primary carbonate content (C_o) or primary noncarbonate content (NC_o), respectively, in fine-grained, calcareous sediments with standardized noncarbonate fractions of NC_d=1 to 15%, and an assumed primary porosity of 80%. K_1=compaction at the onset of cementation, MK=area of mechanical compaction, Z=area of cementation, -Z=area of chemical compaction.
Example: Rock sample with a standardized noncarbonate fraction of 10% (arrows).

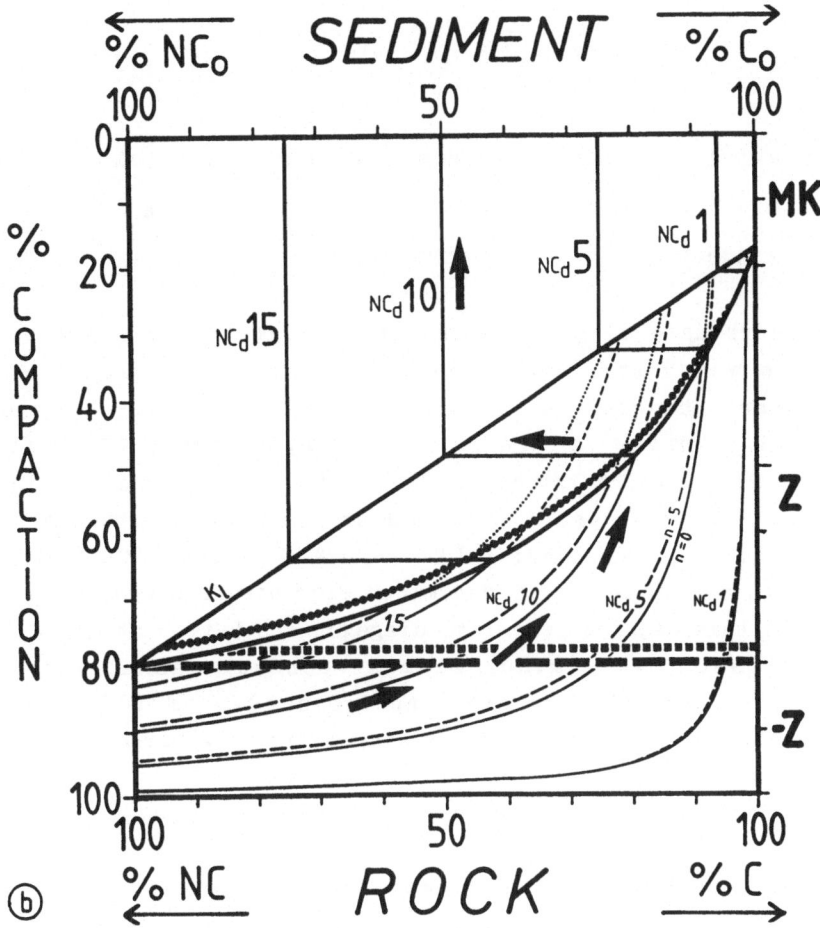

x-axis to the regression curve correlating mechanical compaction with primary carbonate content and then reads the original carbonate content on the upper scale.

7 BURIAL DIAGENESIS OF MINOR
ELEMENTS CONTAINED IN THE
CARBONATE FRACTION

During carbonate diagenesis the concentrations of trace and minor elements which are contained in the carbonate fraction undergo significant changes. As a rule, Mg, Fe, and Mn are considerably enriched, whereas Sr and Na become impoverished (KINSMAN, 1979; VEIZER, 1977; BRAND & VEIZER, 1980). The processes governing these fluctuations occur partly with a relatively low overburden (GIESKES, 1981; BAKER et al., 1982; ELDERFIELD & GIESKES, 1982) while the diagenetic system for major and minor elements is supposed to be still open. In this section, the late diagenetic behavior of minor elements is discussed. This minor element content was affected during carbonate redistribution after an overburden of 100 to 400m was deposited. During this stage of diagenesis, the carbonate system was predominantly closed (see sections 2.3.1 and 3.6).

7.1 Reaction Time and the Problem of Incongruent
Dissolution During Disintegration of the
Carbonate Fraction

The carbonate fraction was dissolved in warm, dilute acetic acid with a pH of 4 to 5. Thereafter, the amounts of Mg, Sr, Fe, Mn, and, to a certain extent, Ca were determined from diluted, residue-free solutions using atomic absorption (HERRMANN, 1975).

During the dissolution of the calcareous fraction, individual components of the residue may dissolve disproportionately. Therefore, reaction time and incongruent dissolution were investigated first. The reaction time was determined by leaching the same sample for differing lengths of time in a 10% acetic acid solution (Fig. 73). Most of the reaction occurred within three hours; thereafter, leaching was low.

However, if one gradually enhances the concentration of the acetic acid solution (during disintegration of the same sample by using different acid concentrations), evidence of incongruent dissolution

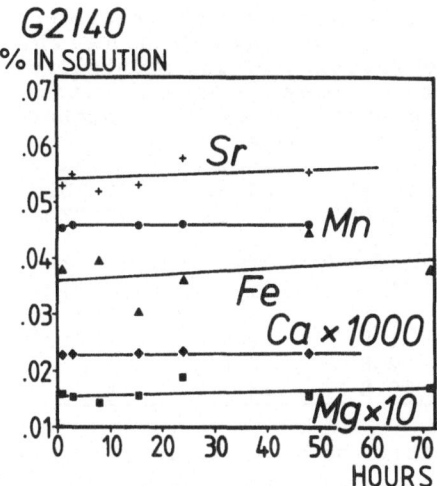

Fig. 73 Leaching of the noncarbonate residue from Gubbio 2 sample 40 (59% CaCO3). The sample consisted of 100mg which was then placed in 5ml of 10% acetic acid solution.

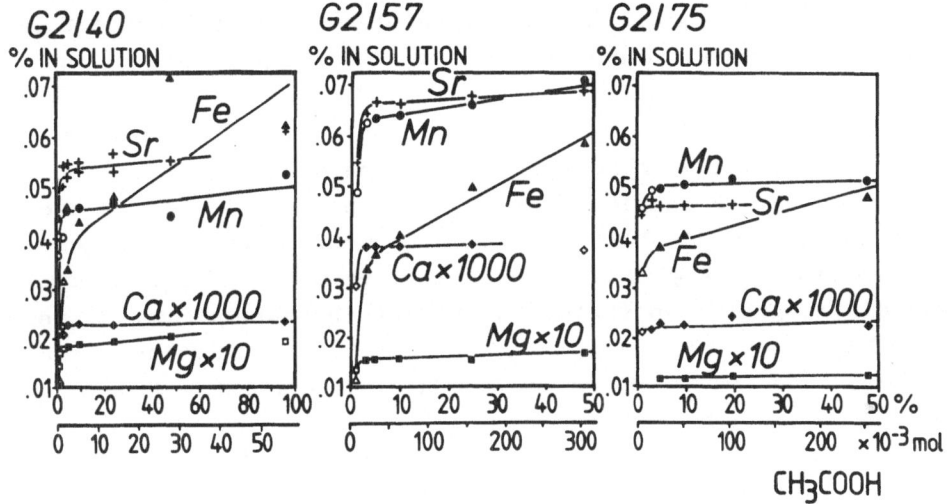

Fig. 74 Incongruent mobilization of Fe after dissolution of the easily soluble carbonate fraction. Samples 40, 57, and 75 (200mg each) from Gubbio 2.

appears (Fig. 74). During the early phase of disintegration (at low acid concentrations), the element concentration quickly increase due to dissolution of the easily soluble carbonate phases. After that (at high acid concentrations), Ca, Sr, Mg, and Mn increase slowly. Their increase is more or less proportional to their total amounts in the dissolved fraction. In contrast, the Fe content rises over-

Table 17: Ratio between the slope coefficient
of the regression curve (describing the
dissolution of the fraction of low solubility)
and the mean minor element concentration
contained in the carbonate fraction. Numbers
are standardized on the Ca value.

Sample	G2 40	G2 57	G2 75
Total carbonate content (%)	59.08	94.00	65.37
Ca	1	1	1
Sr	1.8	1.7	0.9
Mg	1.8	2.4	1.7
Mn	2.8	3.5	1.5
Fe	10.8	15.1	14.0

proportionally. This is shown in Table 17, in which the increase in
concentration in the solution is related to the total concentration of
the measured elements.

The relative composition of the easily soluble (E) and less
soluble (L) phases shows that in both types carbonate is dissolved
(Fig. 75). According to the Ca values, the amount of dissolved pure
$CaCO_3$ is 98 to 99% in the easily soluble phase and 94 to 96% in the
less soluble phase. Despite this, the incongruent mobilization of Fe
is supposedly caused by the leaching of the carbonate-free residue.
Therefore, the concentration of the acetic acid solution was chosen so
that mainly the easily soluble portion would be dissolved. Two
hundred milligram samples from the Gubbio 2 and Neuffen 2 sections
were leached with 20ml of 5% acetic acid solution. However, in
samples from the Angles 2 section a certain amount of sample substance
was disintegrated so that in each case the same amount of residue
(50mg) was in contact with 5ml of 10% acetic acid solution. For this
purpose, the amount of residue was first determined by NaOH titration
(see section 2.1.1).

Numerous measurements of several samples (recording multiple
disintegrations with different standard solutions) furnished an
accuracy for Ca ±4000ppm; for Mg, ±100ppm; for Sr, ±20ppm; for Fe,
±100ppm; and for Mn, ±10ppm. If one compares measurements taken under
the same analytical conditions, the accuracy is improved: Ca ±1500ppm;
for Mg, ±100ppm; for Sr, ±10ppm; for Fe, ±70ppm; and for Mn, ±5ppm.

7.2 Concentrations of Minor Elements Contained

   In the Carbonate Fraction

In the three sections (Gubbio 2, Angles 2, and Neuffen 2) in which

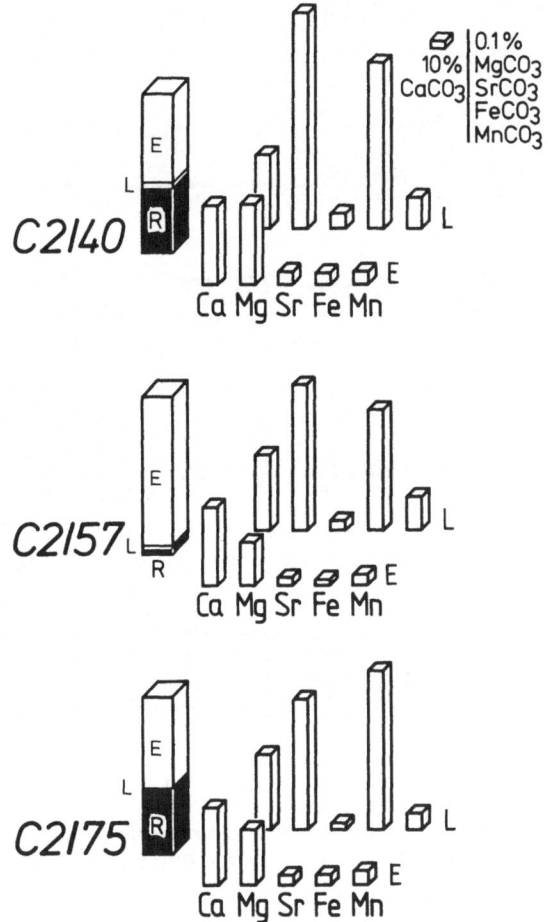

Fig. 75 Proportions of the readily soluble carbonate fraction (E) to that of low solubility (L), and the amount of residue (R) in three samples from the Gubbio 2 section (left column, acetic acid solution). Relative composition of the easily soluble carbonate fraction (E) and that of low solubility (L, right).

minor elements were studied the carbonate fraction contains 0.3 to 1.5 weight % of minor and trace elements (Fig. 76). The three sections exhibit large fluctuations in their minor element concentrations due to both paleogeographic position and the specific manner of diagenesis. For instance, the foraminiferal limestones from Gubbio (deposited in water approximately 1000m deep; ARTHUR, 1976) display comparatively high Mn contents, and very low Fe and Mg contents (see also RENARD, 1979). On the other hand, the alternations of Angles and Neuffen (which were deposited in relatively shallow water) have very

GUBBIO ANGLES NEUFFEN

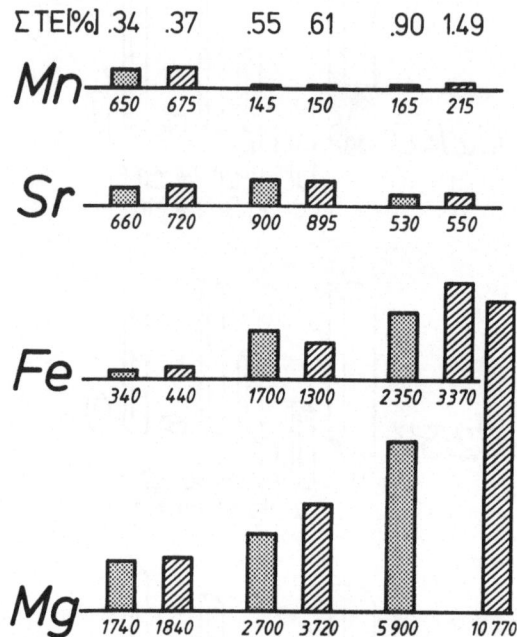

Fig. 76 Mean amounts of minor elements (TE) which are contained within the carbonate fraction (in ppm of the total carbonate content). Concentrations are given for limestone layers (shaded) and marl beds (hatched), respectively.

high Fe and Mg concentrations.

The mean amounts of Sr from Neuffen, Gubbio, and Angles (540, 690, and 900ppm, respectively) are relatively high compared to data given for pre-Tertiary limestones (WEDEPOHL, 1979). In the three sections the amount of Sr is inversely proportional to the total carbonate content (which is 89.5, 84.8, and 76.1%, respectively). Various observations have shown that high clay contents in calcareous sediments partially hamper the diagenetic exchange and removal of Sr from the carbonate phases (KNOBLAUCH, 1963; FLÜGEL & WEDEPOHL, 1967; WEDEPOHL, 1970; VEIZER & DEMOVIC, 1974; KRANZ, 1976; VEIZER, 1977; BAUSCH & POLL, 1984).

The concentrations of Mg, and also of Sr and Mn, although to a somewhat lesser degree, are higher in the marl beds than in the limestone layers; but the amount of Fe is either lowered or elevated in the marl beds of the three investigated sections (Fig. 76). Plotted against the total carbonate content (Figs. 77, 78), the concentration of minor elements exhibits a fairly wide scatter. This

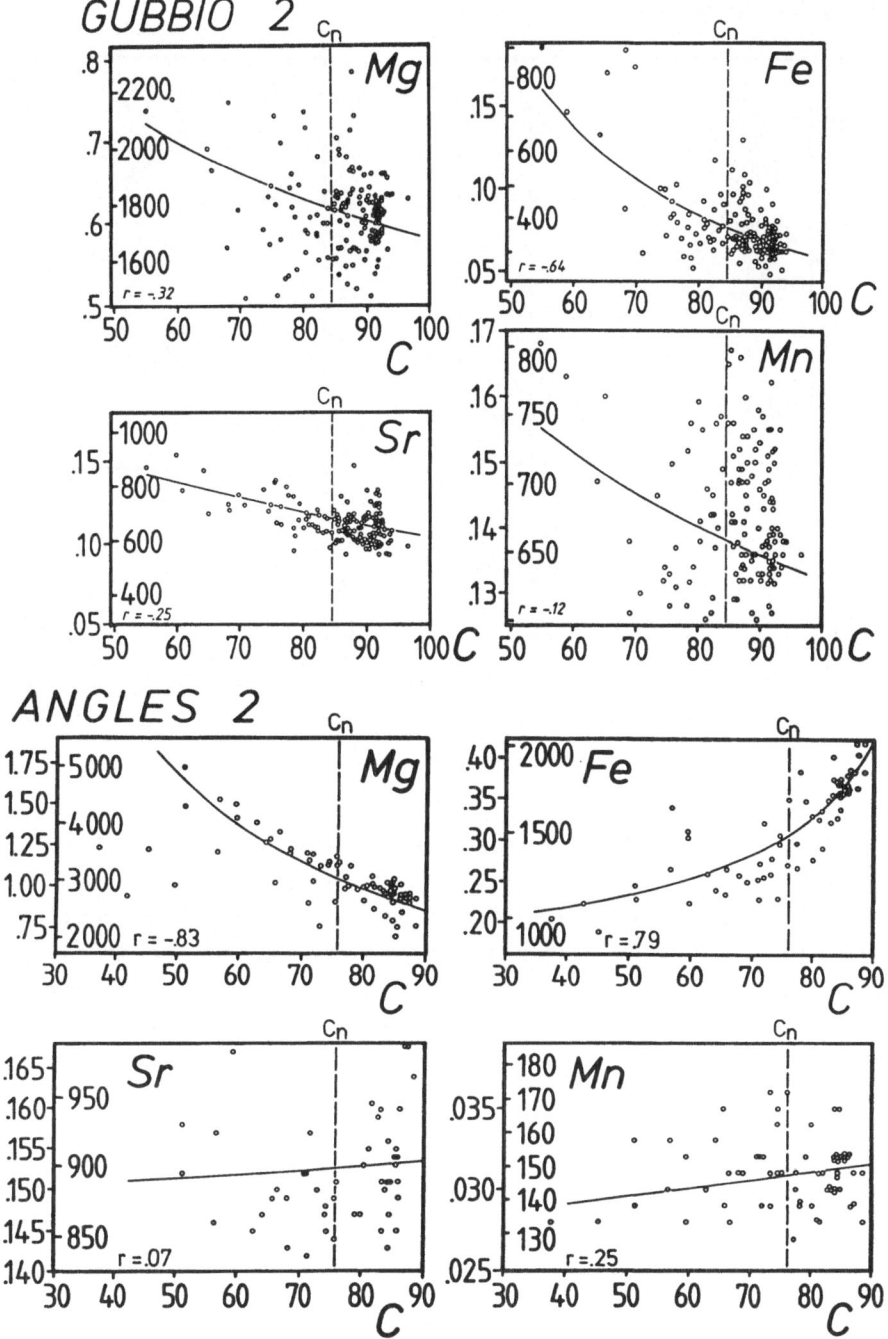

Fig. 77, 78 Relationship between minor elements (TE) which
are contained in the carbonate fraction and the carbonate
content of the rock (C). Vertical axis: concentrations of
the minor elements expressed as percentage $TECO_3$ of the
total carbonate fraction (numbers to the left) and expressed
as ppm TE of the total carbonate fraction (numbers to the
right). Dashed line: carbonate neutral value (C_n).

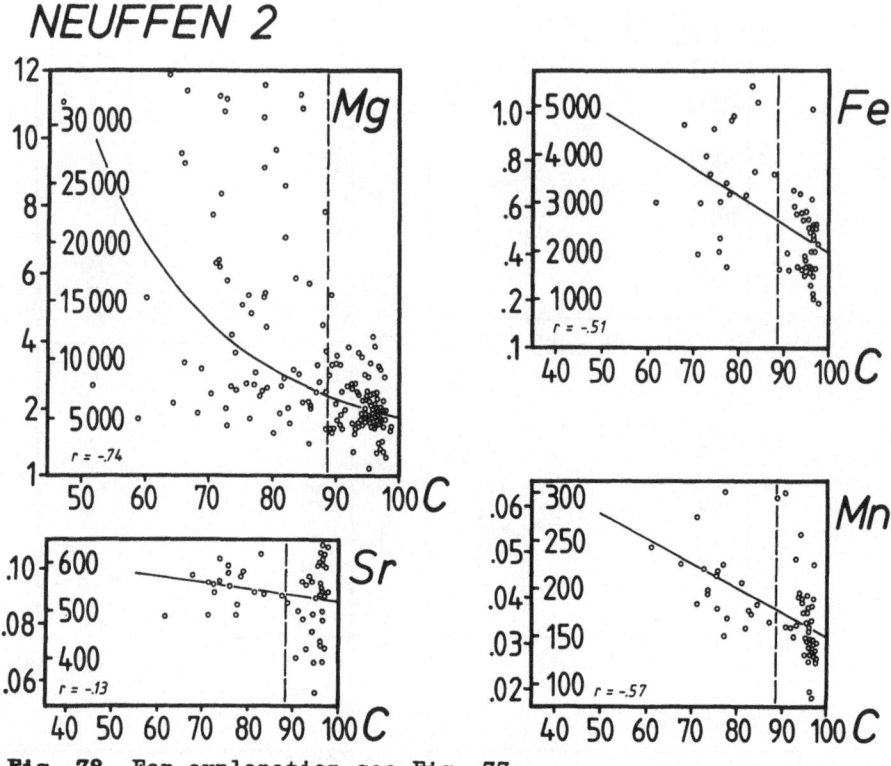

Fig. 78 For explanation see Fig. 77.

is due to analytical errors and to increasing or decreasing
concentration trends within the studied sections (see Figs. 23, 34,
40). According to regression curves between the minor element content
and the amount of total carbonate, maximum differences in the minor
element concentration between the carbonate-rich and the carbonate-
poor layers are as follows: Mg, 25000ppm; Fe, 2500ppm; Sr, 200ppm;
and Mn, 100ppm.

7.3 Interpretation of Minor Element Behavior

7.3.1 Minor Element Mass Balance Calculation

In marl-limestone alternations, four different types of carbonate have
to be distinguished. These are dissolved (-Z) and relic carbonate (R)
in the dissolution zones, and the cemented (Z) and primary carbonate
(P) in the cementation zones. From the bulk concentrations of the
various minor elements contained in the total carbonate fraction, the

minor element content in the four carbonate types can be evaluated by using a mass balance calculation. This depends on the following preconditions:

1. One has to know the mass proportions of the four carbonate types for a given alternation.

2. One also has to know the minor element content of the "primary" carbonate in the cementation zones.

The quantities of and proportions between the above-mentioned carbonate fractions can be determined by performing a carbonate mass balance calculation (see section 2.3), or by using the cement number (z, eq. 21) which is defined as the ratio between the absolute amounts of cemented (or dissolved) carbonate and "primary" carbonate (see section 5.3). Additionally, the minor element content of the "primary" carbonate fraction within the cementation zone has to be evaluated. For reasons of simplification, it is assumed that its minor element concentration was not affected by dissolution and cementation processes during carbonate redistribution. Therefore, the minor element content of the "primary" carbonate fraction is supposed to be equivalent to the minor element content of the neutral value (TE_n). The carbonate neutral value represents the carbonate content at the boundary between the dissolution and cementation zones. It must be stressed, however, that the minor element content of the neutral value does not usually correspond to the original, pre-diagenetic amount of minor elements in the primary sediment. The trace element balance calculation gives only the present, post-diagenetic concentrations of minor and trace elements in the carbonate fractions within the existing rock.

The minor element content of the carbonate cement (TE_z) is a function of the ratio between the absolute values of primary carbonate and cement ($1/z$, eq. 21), the measured bulk trace element concentration (TE, in ppm) and the minor element content of the "primary" carbonate, which is assumed to be equivalent to the trace element content of the carbonate neutral value (TE_n):

$$TE_Z[ppm] = \frac{1}{z}(TE-TE_n)+TE \quad . \tag{28}$$

Eq. 28 has to be manipulated in order to determine the minor element content of the dissolved carbonate (TE_{-Z}) in the dissolution zones. The ratio between the primary carbonate content and the cement content

(1/z) as used for the cementation zones now corresponds to the ratio
between the relic carbonate and the dissolved carbonate $(R/-Z=(1/z)+1)$;
the cement number (z) is negative in the dissolution zones:

$$TE_{-Z}[ppm] = (\frac{1}{z}+1)(TE_n-TE)-TE_n \quad . \tag{29}$$

Using eqs. 28 and 29, the minor element concentrations of the
dissolved and cemented carbonate were determined before being
compared.

A schematic presentation of the method is given in Fig. 79 using
the Fe content from the Gubbio section. The figure displays the
regression curve between total $CaCO_3$ versus carbonate-bounded Fe (from
Fig. 77), and shows the amounts of the four carbonate types. They are
given as histograms which result form the carbonate mass balance
calculation (see Fig. 41c). First, the Fe content of the carbonate
neutral value (C_n) is evaluated (Fe_n=370ppm); and the cement number
(z) is calculated for the different carbonate classes. Then, the
minor element concentration in the four carbonate fractions is
determined by using eqs. 28 and 29. For example, the Fe mass balance
calculation assumes that the "original" carbonate contained 370ppm Fe
throughout the carbonate fraction (prior to the onset of diagenetic
redistribution). However, only 270ppm Fe was dissolved from a portion
of the carbonate contained in the marl layers; this 270ppm Fe was
reprecipitated in the carbonate cement of the cementation zones.
Thus, 100ppm Fe remained in the dissolution zones and was incorporated
into the relic carbonate of those zones. Therefore, the Fe content in
the relic carbonate increased from 370 to 440ppm (according to the
ratio of 0.71 between the amounts of the dissolved and relic carbonate
in the dissolution zones).

Fig. 79 Schematic representation of a minor element mass
balance calculation using the Fe concentration from the
Gubbio 2 section. Above: regression curve of the Fe
content versus total carbonate content (from Fig. 77).
Fe_n=Fe content of the carbonate neutral value. Upper
histogram: dissolved (circles), cemented (crosses),
residual and "primary" carbonate (hatched) for various
levels of $CaCO_3$. Calculation of the "cement number" (z).
Lower histogram: computed Fe content and means for the
dissolved (-Z), cemented (Z), residual (R), and "primary"
(P) carbonate fractions.

MINOR ELEMENT MASS BALANCE

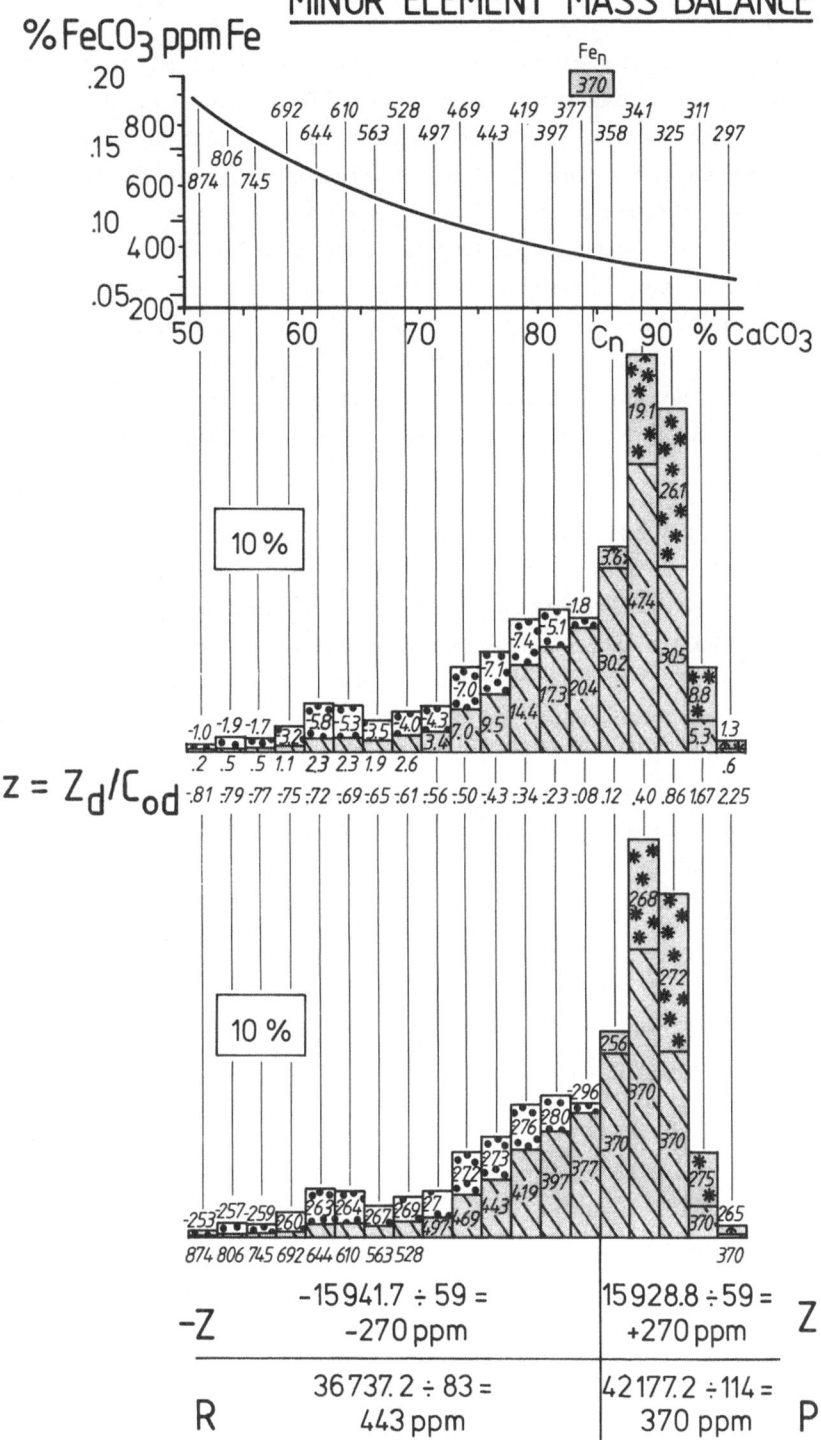

%FeCO₃ ppm Fe

$$z = Z_d/C_{od}$$

-Z   −15 941.7 ÷ 59 =
      −270 ppm

Z   15 928.8 ÷ 59 =
      +270 ppm

R   36 737.2 ÷ 83 =
      443 ppm

P   42 177.2 ÷ 114 =
      370 ppm

Results (Fig. 80):

1. In spite of the low correlations between several minor element and total carbonate amounts (Figs. 77, 78), the calculated minor element content in the dissolved carbonate fraction essentially is equivalent to the calculated minor element content in the cemented carbonate. Consequently, the minor elements released from the carbonate in the marl layers seem to be completely redeposited in the cement of the limestone layers. Thus, during late diagenesis, the marl and limestone layers formed a closed system for carbonate as well as the minor elements. This does not exclude the possibility, however, that during early or very late diagenesis the total amount of trace elements was affected by diagenetic changes.

2. In the residual carbonate in the dissolution zones (within any one section) the various minor and trace elements are enriched or impoverished to varying degrees. Accordingly, the cemented carbonate has a minor element distribution which is in sharp contrast to that of the residual carbonate. For example, compared to the minor element concentration in the cement, Mg increases in the residual carbonate of the marl bed in the three sections by a factor of 1.1 to 2.5. In this fraction, Fe is both considerably reduced and enriched by a factor of 0.6 to 1.8, whereas Sr and Mn are just slightly enriched or diminished in the relic carbonate in the marl beds.

7.3.2 Enrichment of Minor Elements due to Differential Compaction

In principle, one can calculate the diagenetic enrichment of minor elements in the marl beds if one considers those elements as insoluble particles, which are enriched by differential compaction. For this purpose, the basic equation in section 4.4.4 (eq. 12), which relates the existing amount of particles (N) contained in a given volume of rock to a certain degree of compaction (K), has to be rewritten. In this section, the existing amount of particles (N) is expressed as the addition of the number of primary particles (N_O) and of the number of enriched particles (ΔN).

GUBBIO-2

C = 76.7   89.5

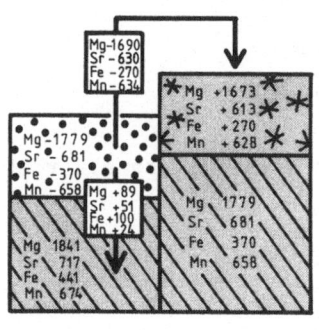

Mg	1841	Mg	1743
Sr	717	Sr	658
Fe	441	Fe	336
Mn	674	Mn	648

ANGLES-2

C = 63.4   83.9

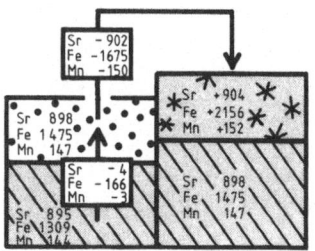

Mg	3722	Mg	2708
Sr	895	Sr	900
Fe	1309	Fe	1707
Mn	144	Mn	149

NEUFFEN-2

C = 77.2   94.0

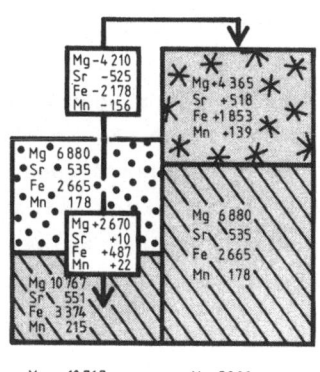

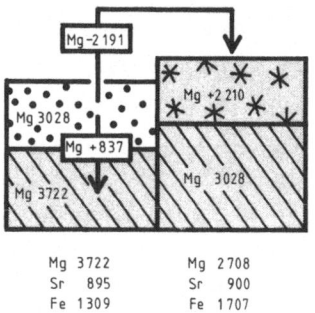

Mg	10 767	Mg	5898
Sr	551	Sr	529
Fe	3374	Fe	2355
Mn	215	Mn	165

Fig. 80 Minor element mass balance calculation for the Gubbio, Angles, and Neuffen sections, respectively. Diagrams show the concentrations of minor elements (in ppm) which are removed from the carbonate in the dissolution zones (left side of the diagrams) and are reprecipitated in the cementation zones (right side). The diagrams also represent the amount of minor elements which are enriched in or depleted from the relic carbonate in the dissolution zones. Symbols are the same as in Fig. 79. C=mean post-diagenetic carbonate content of dissolution and cementation zones. Values below: mean post-diagenetic minor element concentrations (in ppm) measured in the carbonate fraction of both zones.

$$N = N_O + \Delta N \quad . \tag{30}$$

If one substitutes eq. 12 for N, the increasing amount of particles (ΔN) then becomes

$$\Delta N = N - N_O = \frac{N_O K}{100 - K} \quad . \tag{31}$$

Accordingly, the number of particles (N) within a certain rock volume is

$$N = N_O + \frac{N_O K}{100 - K} \quad . \tag{32}$$

On the following pages, eq. 32, which describes the particle enrichment due to compaction, is transformed into an equation which calculates the compactional enrichment of minor elements. First, only the maximum enrichment of minor elements is described assuming that they are perfectly insoluble substances. Contrary to eq. 32, minor elements are enriched only in the carbonate fraction and not within the total sediment volume, as it was previously assumed for particle enrichment. Therefore, the initial volume in eq. 32 which is expressed as "100" percent is now equivalent to the absolute amount of carbonate or the term $(100 - K_n - n_d - NC_d)$. In this term, K_n is the degree of compaction at the carbonate neutral value between dissolution and cementation zones, because the enrichment of minor elements is expressed relative to the minor element content of that value. If one replaces the absolute porosity (n_d) by eq. 15 (section 5.3), the starting carbonate volume for the maximum enrichment of minor elements is given by:

$$(100 - K_n - NC_d + n(0.01 K_n - 1)) \quad . \tag{33}$$

Now, eq. 32 can be transformed to deal with the enrichment of minor elements contained within the carbonate fraction. The amounts of the existing and original particles (N and N_O, respectively, eq. 32) correspond to the measured minor element content of the dissolution zone (TE_{-z}) and to the concentration of minor elements at the carbonate neutral value (TE_n). The primary volume used in the calculations of grain enrichment (which is expressed by the number

"100" in eq. 32) must be replaced by term 33. Moreover, compaction (K) is expressed as the increasing amount of compaction relative to the compaction at the neutral value ($K-K_n$). This is because enrichment in the marl beds only occurs when compaction at the neutral value is exceeded. Hence, the maximum minor element enrichment (TE_{-Zmax}) is expressed as the addition of the initial concentration and of the maximum enriched concentration of minor elements ($TE_{-Zmax}=TE_n+\Delta TE_{max}$):

$$TE_{-Zmax}[ppm] = TE_n + \frac{TE_n(K-K_n)}{100-NC_d-K+n(0.01K_n-1)} \quad . \quad (34)$$

In naturally occurring alternations, however, the compactional enrichment of minor elements is largely incomplete, since the studied minor elements are not completely insoluble, as is required for particle enrichment (section 4.4.4). In order to quantify the completeness of the enrichment processes, the term "closure" (X) is introduced, which is defined as the ratio between the actual increase (ΔTE) and the maximum possible increase of minor elements during compaction (ΔTE_{max}):

$$X[\%] = \frac{100\Delta TE}{\Delta TE_{max}} \quad . \quad (35)$$

For example, if marl layers are a completely closed system (X=100%) and minor elements behave like insoluble particles, the actual increase would equal the maximum increase; hence chemical compaction causes a complete enrichment of minor elements. On the contrary, at X=0%, no enrichment relative to the initial value (TE_n) occurs, while at negative X values the minor element concentration is impoverished.

According to eq. 35 the actual increase (ΔTE) is equivalent to the expression $\Delta TE=\Delta TE_{max}X/100$. In this equation, the maximum increase (ΔTE_{max}) has to be substituted by its similar expression ($TE_n(K-K_n))/(100-NC_d-K+n(0.01K_n-1))$ in eq. 34. Finally, the compactional enrichment of minor elements within the marl bed is the addition of the intitial concentration (TE_n) and of the actual increase (ΔTE):

$$TE_{-Z}[ppm] = TE_n + \frac{TE_n(K-K_n)}{100-NC_d-K+n(0.01K_n-1)} 0.01X \quad . \quad (36)$$

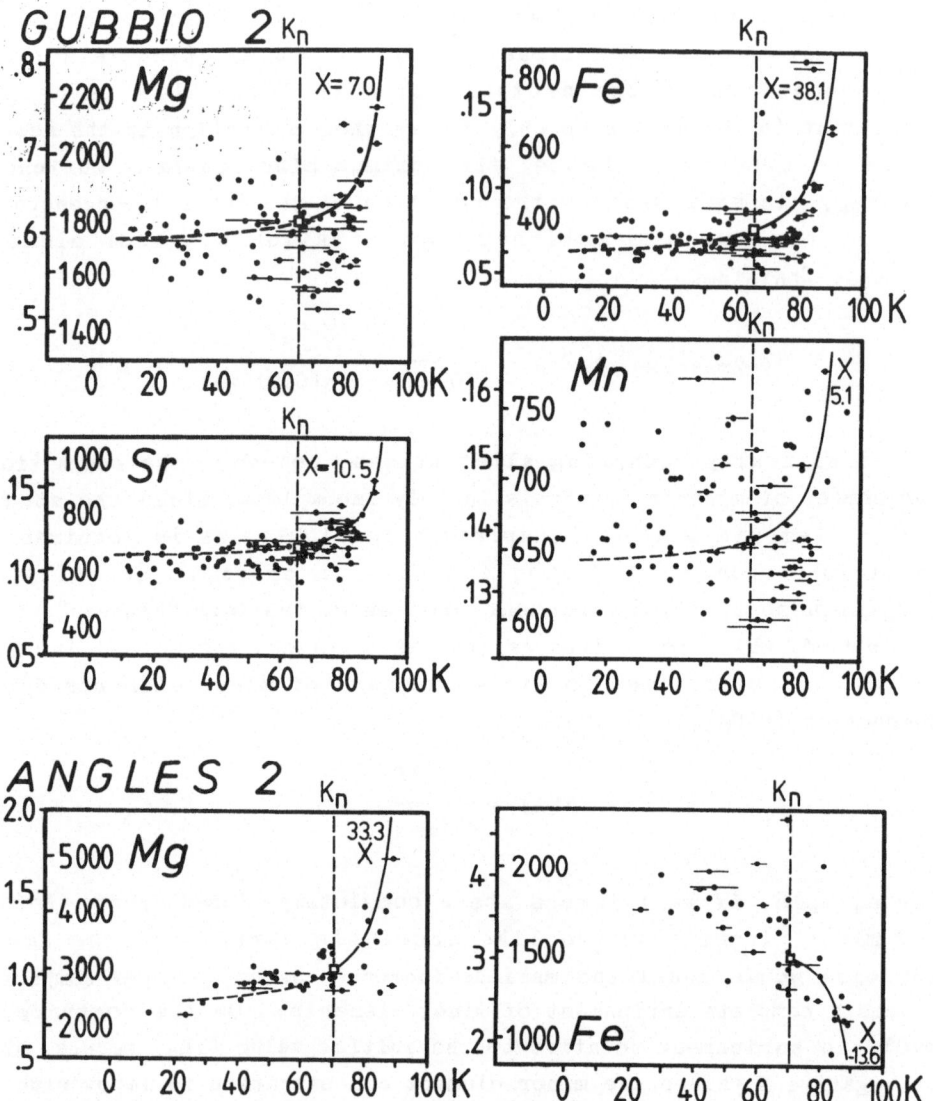

Fig. 81, 82 Enrichment of minor elements, which are contained within the carbonate fraction, via compaction (K). Vertical axis: minor element concentration (TE) expressed as percentage TECO$_3$ of the total carbonate fraction (numbers to the left), and as ppm TE of the total carbonate fraction (numbers to the right). Curves: theoretical enrichment due to compaction as calculated by eq. 36 with values of closure (X) in the system of the dissolution zones equivalent to the influx or outflux of minor elements as displayed in Table 18. K$_n$=compaction at the carbonate neutral value.

The amount of inclusion of the various minor elements in the marl beds is determined by using eq. 35 and the results of the minor element mass balance calculation (section 7.3.1). The determination

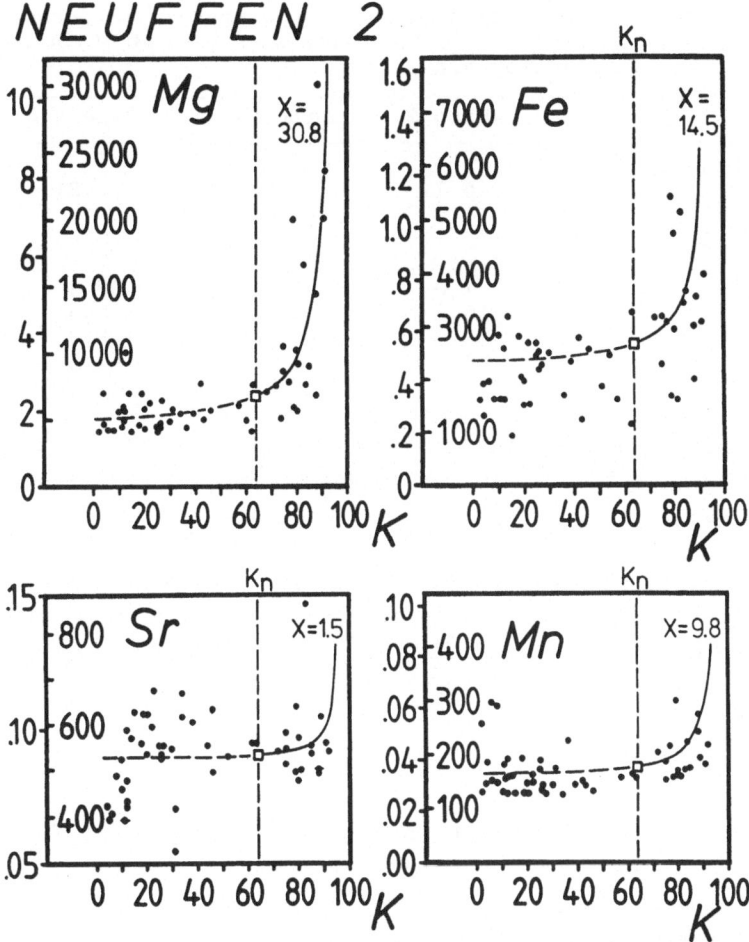

shows (Table 18) that enrichment or impoverishment is different for each of the various elements within a single section. According to PINGITORE (1982), this is a typical phenomenon of trace element diagenesis. In the three sections, the minimum and maximum closure of the minor elements varies from -14 to 38%.

The enrichment of Mg, Sr, Fe, and Mn in the carbonate fraction of the marl beds is calculated by using eq. 36 and the amount of closure from Table 18. There is a clear correspondence between the calculations and the measurements correlating minor element content with compaction (Figs. 81 and 82). Both calculated curves and measured data show that the minor element content in the carbonate fraction is enhanced as much as six times relative to that at the carbonate neutral value. For example, in the Neuffen section the Mg content increases from 5000 to 30,000ppm (with 90% compaction); this equals about 10 weight % $MgCO_3$ or about 22 weight % dolomite. Several authors already described burial dolomitization of stylolitic seams

Table 18: Mean amounts of closure in the system (in percent) for the different minor elements contained in the carbonate fraction of the dissolution zones.

Section	Gubbio 2	Neuffen 2	Angels 2
Ratio between the amounts of dissolved (-Z) and relic carbonate (R)	0.71	1.26	0.81
Mg	7.0	30.8	33.3
Sr	10.5	1.5	-0.5
Fe	38.1	14.5	-13.6
Mn	5.1	9.8	-2.5

and wavy pressure dissolution seams (BERGER & RAD, 1972; WANLESS, 1979; MATTES & MOUNTJOY, 1980; JØRGENSEN, 1983). In principle, there are two possibilities to explain this phenomenon.

1. Accumulation of pre-existing dolomite crystals by the preferential dissolution of $CaCO_3$ (BERGER & RAD, 1972; MATTES & MOUNTJOY, 1980; JØRGENSEN, 1983).

2. New formation of dolomite (WANLESS, 1979; MATTES & MOUNTJOY, 1980). For instance, the predominant dissolution of Mg-calcite could increase the Mg/Ca ratio in the pore solution of the marl beds which favors dolomitization (LIPPMANN, 1973; FOLK & LAND, 1975). Moreover, MATTES & MOUNTJOY (1980) explain the dolomitization of pressure dissolution seams as Ca/Mg exchange at the strained surface of the calcite crystals.

Unlike the behavior of other elements, the Fe content, which exhibits enrichment or depletion in the cement of various sections, presumably results from the absence or existence of sulfate reduction processes. If sulfate is already present in the pore water, Fe is consumed by sulfide precipitation and leaves the carbonate system. Then, after the consumption of sulfate, the Fe^{2+} content again increases in the pore waters if, as is usually the case, reducing conditions are maintained (FÜCHTBAUER & MÜLLER, 1977; CURTIS, 1977; BERNER, 1980; 1984). The zone of sulfate reduction, which often begins within or closely below the sediment-water interface, may extend a few to several 100 meters downward (GIESKES, 1981). The carbonate cements of the Gubbio and Neuffen sections display relatively low Fe concentrations. The limestones had relatively low mechanical compaction and low overburden (80 and 140m) when cementation set in; thus cementation presumably occurred during sulfate reduction. On the other hand, in the Angles section cementation began with a higher overburden (170m), and sulfate reduction was supposedly completed since the cements are enriched in Fe.

7.3.3 Composition of the Pore Solution

The concentration of the minor elements in the various types of carbonate (e.g., "primary" and cemented carbonate; section 7.3.1) enables one to estimate the composition of the ancient pore waters for different values of sedimentary overburden using distribution coefficients. The distribution coefficient k_{TE}^{CC} defines the molar ratio for a minor element and calcium (mol TE/mol Ca) between the (pore) solution (s) and the precipitating calcite phase (cc). Accordingly, the relative composition of the pore solution is:

$$molTE_s/molCa_s = \frac{molTE_{cc}/molCa_{cc}}{k_{TE}^{cc}} \qquad . \qquad (37)$$

The distribution coefficient for Fe and Mn, which is usually determined experimentally, still displays considerable uncertainties. New determinations of the distribution coefficient for Sr and Mg vary from 0.035 to 0.05 for Sr (KATZ et al., 1972; BRAND & VEIZER, 1980; BAKER et al., 1982) and between 0.02 and 0.06 for Mg (BRAND & VEIZER, 1980). In this study, calculations were made using values of $k_{Sr}^{CC}=0.035$ (BAKER et al., 1982) and $k_{Mg}^{CC}=0.02$ (MICHARD, 1971; MUCCI & MORSE, 1983). An estimation of the Mg/Ca and Sr/Ca molar ratios in the pore water (eq. 37) is tentatively calculated for the following three stages which represent increasing diagenesis (Table 19):

1. Early Stage: The highest pore water molar ratio (Mg/Ca, Sr/Ca) is described by the minor element content of the carbonate neutral value. As previously discussed (section 7.3.1), the minor element content of the neutral value is equivalent to that of the "primary" carbonate. If no alteration of the minor element concentration in the "primary" carbonate fraction occurred, its concentration corresponds to a relatively early diagenetic stage which is equivalent to the onset of cementation. Additionally, one must assume that the minor element content of the neutral value was in equilibrium with the surrounding pore water. Both assumptions, however, may not be valid.

2. <u>Intermediate Stage</u>: The pore water element ratios were calculated using the average minor element content of the carbonate cement precipitated in the cementation zones. The minor element content was obtained from the minor element mass balance calculation (section 7.3.1).

3. <u>Late Stage</u>: This stage describes the minor element composition in the carbonate released during the greatest compaction in the marl layers; this carbonate was subsequently reprecipitated as cement in the limestone layers.

Table 19:   Trace and minor elements (in ppm) contained in the carbonate fraction and the relative composition of the pore fluid calculated by using eq. 37.

Stages of dia-genesis*)	Gubbio 2			Neuffen 2			Angles 2		
	1	2	3	1	2	3	1	2	3
Ca **)	396645	396947	396905	388157	392325	393309	394276	394842	395410
Mg	1779	1673	1696	6880	4365	3407	3028	2210	2162
Sr	681	613	639	535	518	528	898	904	900
Fe	370	270	253	2665	1853	2296	1475	2156	1557
Mn	658	628	635	178	193	162	147	152	149
$\dfrac{\text{mol Mg}_S}{\text{mol Ca}_S}$	0.37	0.35	0.35	1.46	0.92	0.71	0.63	0.48	0.45
$\dfrac{\text{mol Sr}_S}{\text{mol Ca}_S}$ $\times 10^{-2}$	2.24	2.02	2.10	1.80	1.73	1.75	2.98	3.13	2.97

*)stages of diagenesis 1 to 3 correspond to an early, medium, and late diagenetic stage, respectively.
**)calculated.

The calculated composition of the pore solution (Fig. 83b) shows only a moderate correlation with the actual curves of Mg/Ca and Sr/Ca molar ratios (Fig. 83a), which are known from both marine, pelagic carbonates (NEUGEBAUER, 1974; SALES & MANHEIM, 1975; GIESKES, 1975; 1981; BAKER et al., 1982) and formation waters in calcareous rocks (ENGELHARD, 1972). This indicates that the molar proportions of the pore water which result from the calculations mostly represent conditions found during late diagenesis and in sections with large thicknesses of overburden. Therefore, as has already been established in simulation models (section 5.2.2), the formation of diagenetic bedding occurs mainly during the late stages of diagenesis.

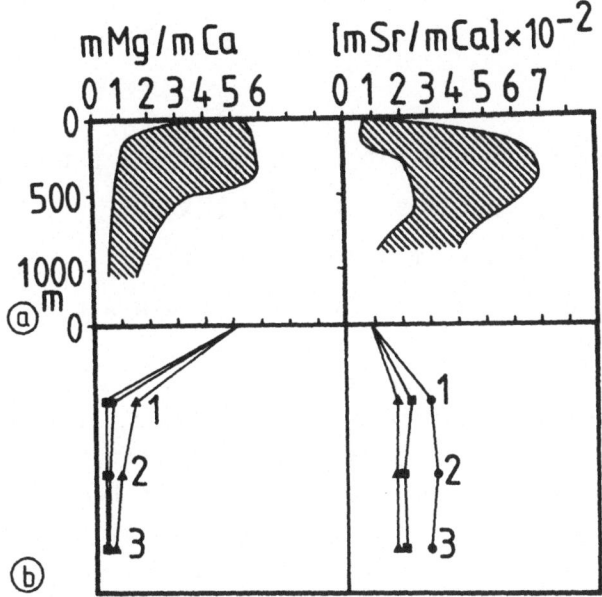

Fig. 83 a: Molar proportions (Mg/Ca; Sr/Ca) of the pore water during increasing overburden of carbonates from the Deep Sea Drilling Project. Data from NEUGEBAUER, 1974; SALES & MANHEIM, 1975; GIESKES, 1975; 1981; and BAKER et al., 1982.
b: Calculated relative composition of the ancient pore water in the Gubbio 2 (■), Neuffen 2 (▲), and Angles 2 (●) sections, respectively. The increased overburden in the diagenetic stages (1 to 3) is not shown to scale. Primary molar proportions of the sea water is Mg/Ca=5.2, and Sr/Ca= 0.86×10^{-2} (KINSMAN, 1969).

7.4 Conclusions

The processes of diagenetic bedding increase not only the amplitude of primary carbonate oscillations, but they also cause differences in concentration of minor elements contained in the carbonate fractions of marl and limestone layers. As minor element mass balance calculations show, this is due to the different concentrations of minor elements in the various carbonate fractions, especially in the cemented and relic carbonate. The most important of the diagenetic processes is the enrichment of Mg and that of several other minor elements in the partly closed marl layers (dissolution zones) due to chemical compaction. The minor elements released from the marl layers were completely reprecipitated in the cementation zones. The calculated relative pore water composition provides further evidence that diagenetic bedding forms mainly during late diagenesis.

8 PRIMARY DEPOSITION OF MARL-LIMESTONE ALTERNATIONS: EXAMPLE FROM THE UPPER JURASSIC, SOUTHERN GERMANY

In lithified marl-limestone alternations, the original mode of deposition is often difficult to determine. First, it is not possible to infer conclusively from the existence of present rhythmic bedding that the primary sediment was actually cyclic because diagenesis considerably enhances the rhythmicity (see section 4.5). Second, primary depositional structures are mostly destroyed due to bioturbation and the development of diagenetic dissolution seams (see Fig. 30H). The well-bedded, Upper Jurassic limestones in southern Germany provide a striking example of these problems. The alternations were thought to be generated by climatic cycles, as evidenced by their present, diagenetically enhanced rhythmicity (e.g., SEIBOLD, 1952; WEILER, 1957; HILLER, 1964; GYGI, 1966; FREYBERG, 1966; KÖHLER, 1971; GWINNER, 1976; BAUSCH et al., 1982; EINSELE, 1982). However, new results show that some portions of the primary stratification was caused by depositional events. A detailed description of the depositional history is given by RICKEN (1985a); therefore, this section presents only a short overview.

The Oxfordian and Kimmeridgian marl-limestone alternations (Fig. 84a) were deposited on the northern shelf of the Tethys Ocean. Presumably, the alternations belonged to a major facies zone which stretched from east to west parallel to the ancient shoreline in the north. This facies zone was probably composed of both an extended, but presently mostly eroded algae-sponge reef belt (GWINNER, 1976) and, towards the basin, marl-limestone alternations. In the northwestern outcrop area of the Swabian Alb, the alternations

Fig. 84 a: Outcrop of the Oxfordian and Kimmeridgian in the Swabian Alb (southern Germany) with the locations of measured sections shown in Fig. 85. Current roses: bipolar measurements of belemnite shells (white) and foreset orientation of ripple laminations in turbidite to tempestite sequences (black). Inset map shows the paleogeography and the supposed flow regime.
b: Cross section through the Swabian Alb (parallel to strike). Crosshatched: biostrome complexes, stippled: marl-limestone alternations in biostrome talus and erosion faces, shaded: micritic marl-limestone alternations.

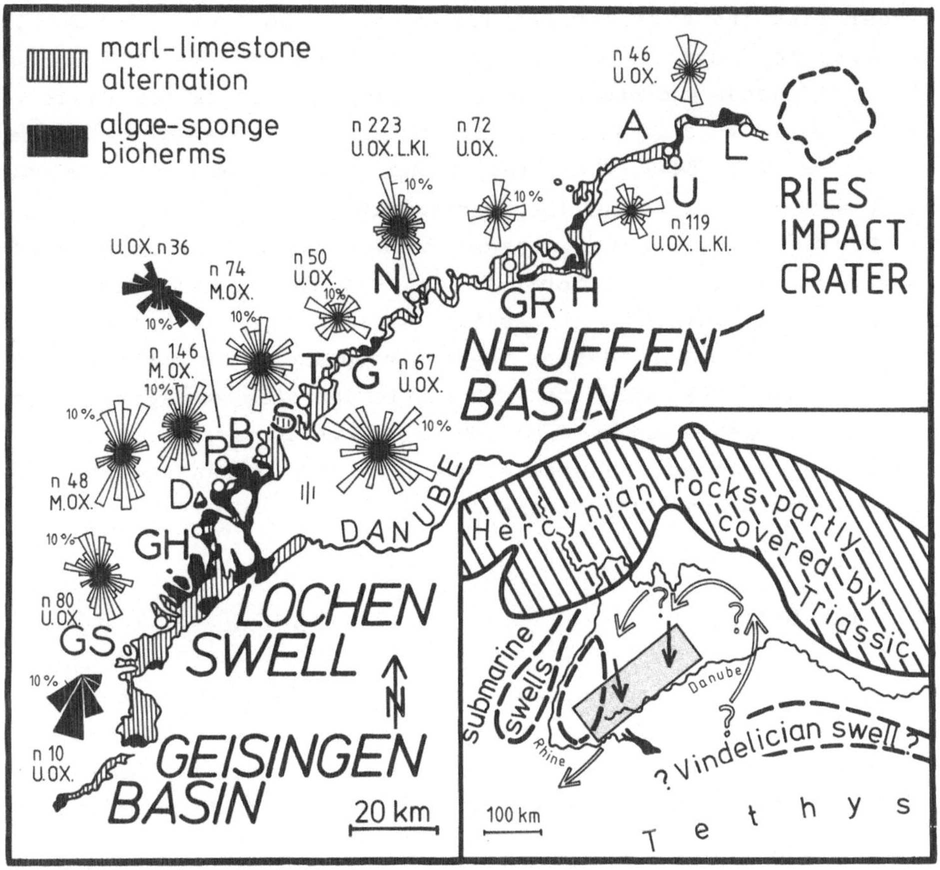

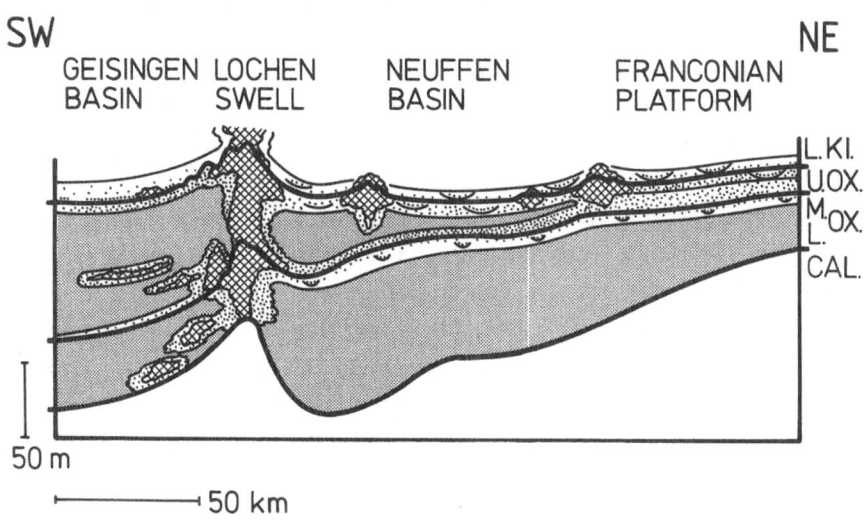

marl–limestone
alternation

algae–sponge
bioherms

n 46
U. OX.

A
L
U

RIES
IMPACT
CRATER

n 223
U. OX. L. Kl.

n 72
U. OX.

10 %

10 %

n 119
U. OX. L. Kl.

U. OX. n 36

n 74
M. OX.

n 50
U. OX.

N

GR H

10 %

10 %

10 %

10 %

NEUFFEN
BASIN

n 146
M. OX.

T
G
S

n 67
U. OX.

10 %

10 %

10 %

P B

D

I ı ı

10 %

n 48
M. OX.

GH

DANUBE

10 %

LOCHEN
SWELL

N

n 80
U. OX.

GS

20 km

GEISINGEN
BASIN

10 %

n 10
U. OX.

Hercynian rocks partly covered by Triassic

Submarine swells

Danube

Rhine

? Vindelician swell ?

100 km

T e t h y s

(a)

SW

NE

GEISINGEN
BASIN

LOCHEN
SWELL

NEUFFEN
BASIN

FRANCONIAN
PLATFORM

L. Kl.
U. OX.
M. OX.
L.
CAL.

50 m

(b)

50 km

interfinger with algae-sponge reefs which were at least 30 to 50m higher than the adjacent sea-floor (Fig. 84b). A prominent, N-S striking reef-swell (Lochen Swell) divided the sea into two basins (i.e., Geisingen and Neuffen Basin). Since major facies differences in the basins were in an N-S direction (see the orientation of current indicators, Fig. 84a), the alternations look very monotonous along the NE-SW striking outcrop of the Swabian Alb.

During transgressive phases (caused by eustasy or by tectonic movements of the total South German Platform), very fine-grained, calcareous turbidites to tempestites and hemipelagic carbonates were deposited. However, during regressive phases, the carbonate content decreased (EINSELE, 1985) and bioclastic detritus was spread over the basins of the Swabian Alb. Regressive phases occurred at the boundaries between the Middle and Upper Oxfordian and between the Upper Oxfordian and Kimmeridgian.

8.1 The Bedding

In outcrops, the bedding exhibits the typical characteristics of diagenetic bedding: Equally thick limestone layers (15 to 30cm thick) alternate with differently thick marl beds. When the mean carbonate content ranges from 75 to 80% (i.e., Lower and Middle Oxfordian and Lower Kimmeridgian), the marl beds are 0.1 to 5m thick (see Fig. 36A). Only when the average carbonate content is high (approximately 90%) do brick-like alternations appear. Then, the marl beds are reduced to only small marl joints (Upper Oxfordian, see Fig. 36B). The bed-by-bed correlation between the various larger outcrops along the Swabian Alb (Fig. 86) provides the following results:

1. The Upper Oxfordian limestones (composed of the marl-limestone alternation Type III) show a clear relationship between the total thickness of the section and the number of bedding planes (Fig. 85). The number of bedding planes (or limestone layers) increases constantly with increasing thickness of the total section. The base and the top of the Upper Oxfordian alternation are actually isochronous boundaries because near the boundaries contemporaneous event layers can be traced laterally throughout the Swabian Alb (Fig. 86). Therefore, new limestone layers of more or less the same thickness are inserted when the total thickness increases. Thus, the Upper Oxfordian marl-limestone alternation is regarded

predominantly as a non-cyclic alternation.

2. The two basins of the Swabian Alb display distinct bedding rhythmicities, although they have both a parallel facies development and the same mean limestone layer thickness. Similar conditions are also found in the subbasins of the Franconian Alb. Every basin shows its own bedding rhythm (FREYBERG, 1966).

3. According to the mode of deposition, distinct, individual beds can be traced for varying distances. In the turbidite to tempestite sequences found in the Geisingen Basin, individual layers pinch out after a few to tens of kilometers. On the other hand, the bedding is more constant in regressive phases, except in channel areas. Marl-rich bedding planes originally caused by lag deposits can be traced 10 to 100km (especially at the base of the Upper Oxfordian).

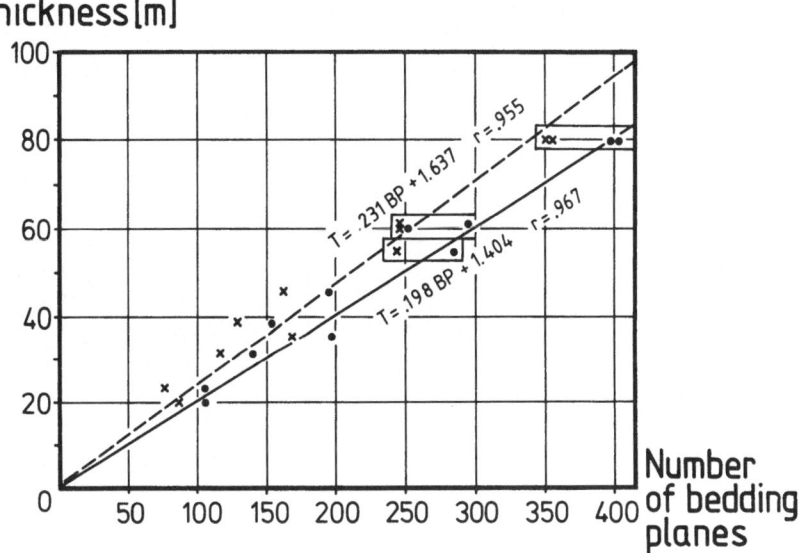

Fig. 85 Thickness of the Upper Oxfordian marl-limestone alternation and number of bedding planes. Swabian Alb, southern Germany. Major bedding plans (x), major and minor bedding planes (●). Bordered data points: number of bedding planes is interpolated by studying approximately one half of the total section and using the entire strata thickness from ZIEGLER, 1977.

At the Oxfordian-Kimmeridgian boundary, the layers systematically pinch out towards the Franconian Platform and the Middle German Land Mass as a result of erosion (FREYBERG, 1966). On the other hand, in

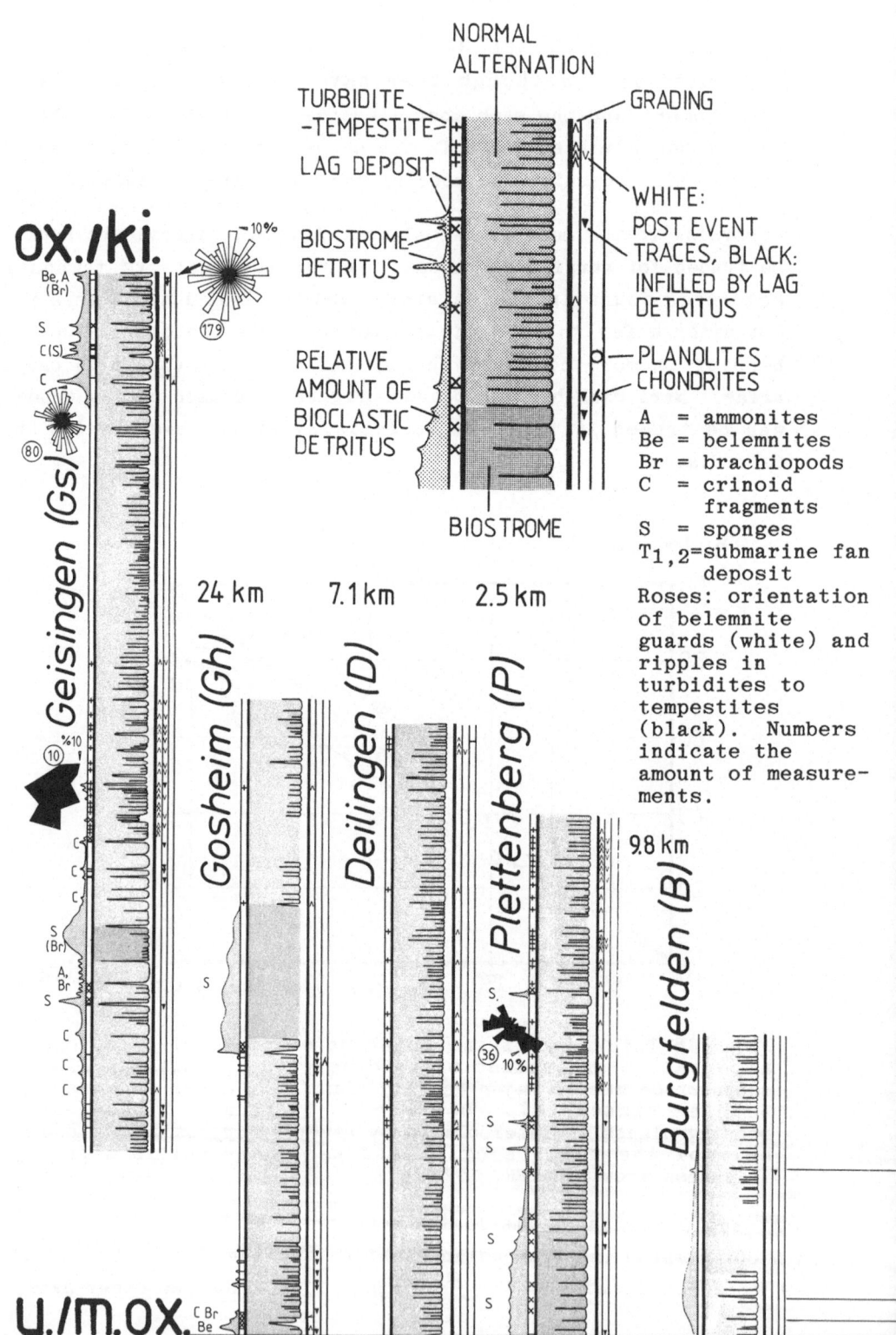

NORMAL
ALTERNATION

TURBIDITE
-TEMPESTITE

GRADING

LAG DEPOSIT

WHITE:
POST EVENT
TRACES, BLACK:
INFILLED BY LAG
DETRITUS

BIOSTROME
DETRITUS

RELATIVE
AMOUNT OF
BIOCLASTIC
DETRITUS

PLANOLITES
CHONDRITES

A = ammonites
Be = belemnites
Br = brachiopods
C = crinoid
     fragments
S = sponges
$T_{1,2}$=submarine fan
     deposit
Roses: orientation
of belemnite
guards (white) and
ripples in
turbidites to
tempestites
(black). Numbers
indicate the
amount of measure-
ments.

BIOSTROME

ox./ki.

Be, A
(Br)

S

C(S)

C

Geisingen (Gs)

%10

C
C
C

S
(Br)

A,
Br
S

C
C
C

24 km

Gosheim (Gh)

S

7.1 km

Deilingen (D)

2.5 km

Plettenberg (P)

S.

10 %

S
S

S

S

9.8 km

Burgfelden (B)

u./m.ox.
C Br
Be

Fig. 86 Bed by bed correlation and sedimentological observations in sections from the Swabian Alb, southern Germany. For location of sections see Fig. 84a.

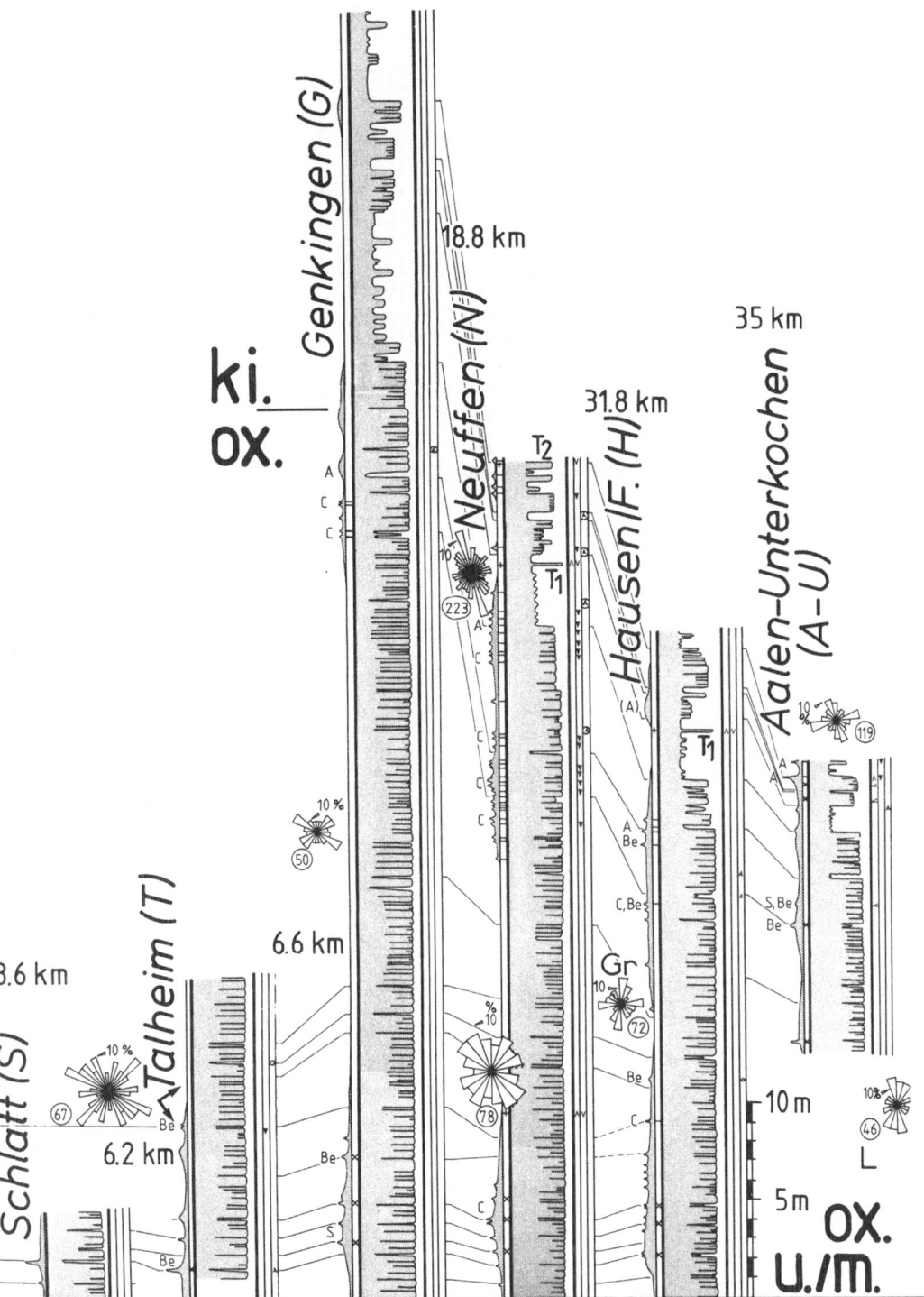

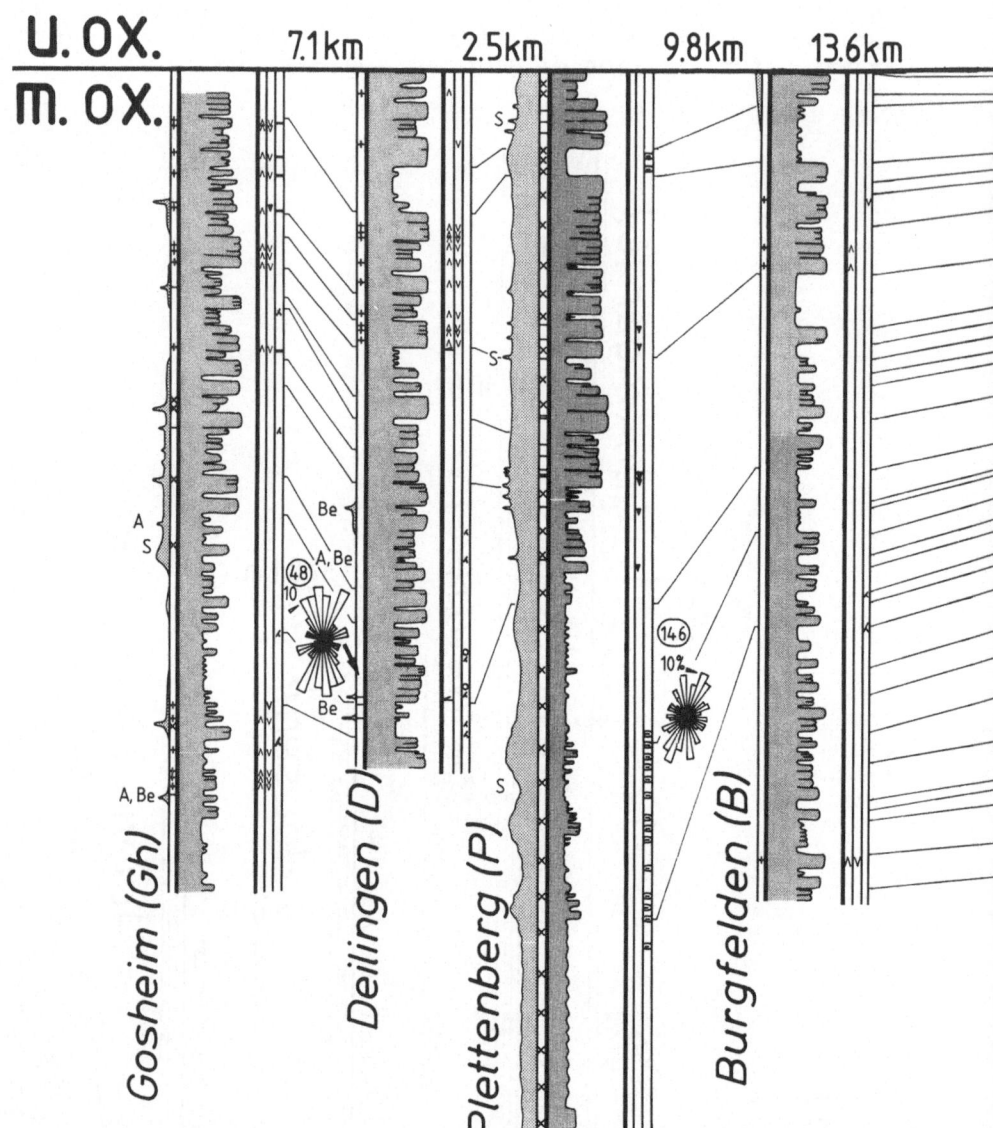

the Lower and Middle Oxfordian in the Neuffen Basin the bedding remains constant for more than 150km. The layers can be traced through the Neuffen Basin over the Ries Meteorite Crater to Franconia (SCHMIDT-KALER, 1962). The cause of this phenomenon remains unclear, because there is little evidence indicating the nature of the primary depositional mode.

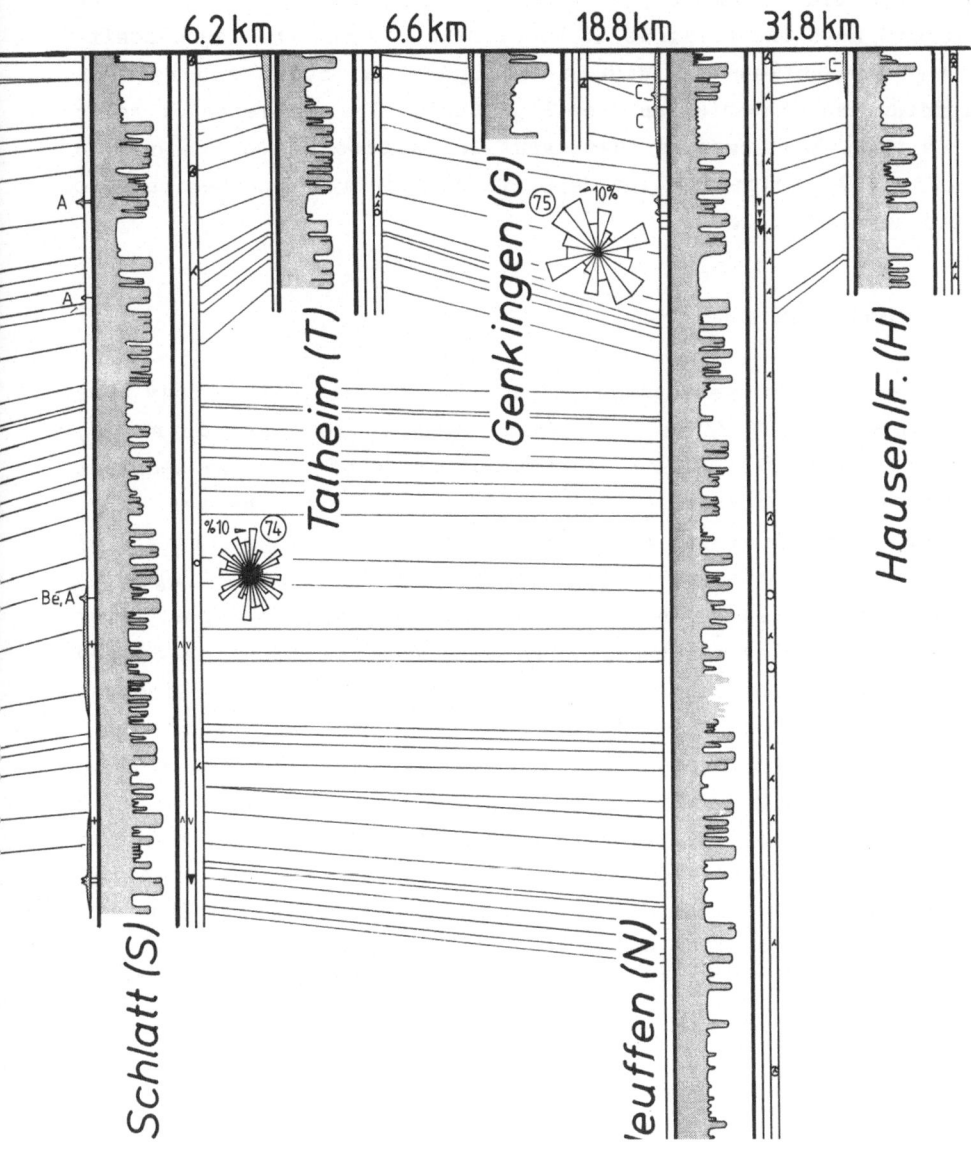

8.2 Processes of Deposition

Relic sedimentary structures in the alternations comprise only a few percent of the present sequence. However, they allow the identification of three major types of event deposition: sedimentary lags, submarine channels and fans, and fine-grained tempestites to turbidites. Due to the presence of two transgressive-regressive cycles in the Middle and Upper Oxfordian sequence, the micritic alternations have two horizons of bioclastic channel and lag deposits (at the boundary between the Middle and Upper Oxfordian and between the Oxfordian and Kimmeridgian, see Fig. 84b). The existence of relic structures (Fig. 85) suggests that event deposition was quite common in the original sediment. The deposition of pelagic and hemipelagic carbonates remains unclear.

8.2.1 Lag Deposits

Bioclastic marl beds are interpreted as lag deposits, since they pass laterally into channel fills and often contain erosional detritus, intraclasts, and shells (Fig. 87). Usually, some of this lag sediment is brought into the underlying limestone layers by bioturbation (Fig. 88A,B). Except in submarine channels, the bioclastic lags consists of parautochthonal assemblages, e.g., articulated crinoid stem fragments, ammonites, and belemnites. Apparently, during short phases of submarine erosion, clay was transported into the basin, which caused the lags to become one of the preferred sites of diagenetic carbonate dissolution. One can use the mean primary carbonate content calculated for dissolution and cementation zones (see section 3.4) to estimate the amount of clay dilution. The average factor of clay dilution in the relatively carbonate poor beds of the primary sediment is 1.3 (Middle Oxfordian) and 1.4 (Upper Oxfordian). If one uses the maximum carbonate variations which are about two times higher than the mean primary values for both zones (see Fig. 14), the factor of clay dilution would be 1.7 and 2.0 for the Middle and Upper Oxfordian, respectively.

8.2.2 Shelf Channel and Fan Systems

Most of the eroded sediment was transported in submarine channels to the south; the sediment was thereby brought into suspension. On the deeper shelf the eroded material was redeposited in submarine fans which were 10's of km wide (Fig. 89). The shelf channels were up to 5m deep and had a maximum width of 50m. Their length is unknown. In outcrop, channels shift relatively slightly in the lateral and in the vertical dimension; thereby causing prograding point bar structures (Fig. 90). Large channels remained active for several million years. The channel fill consists of ammonites, crinoid stem fragments, partly

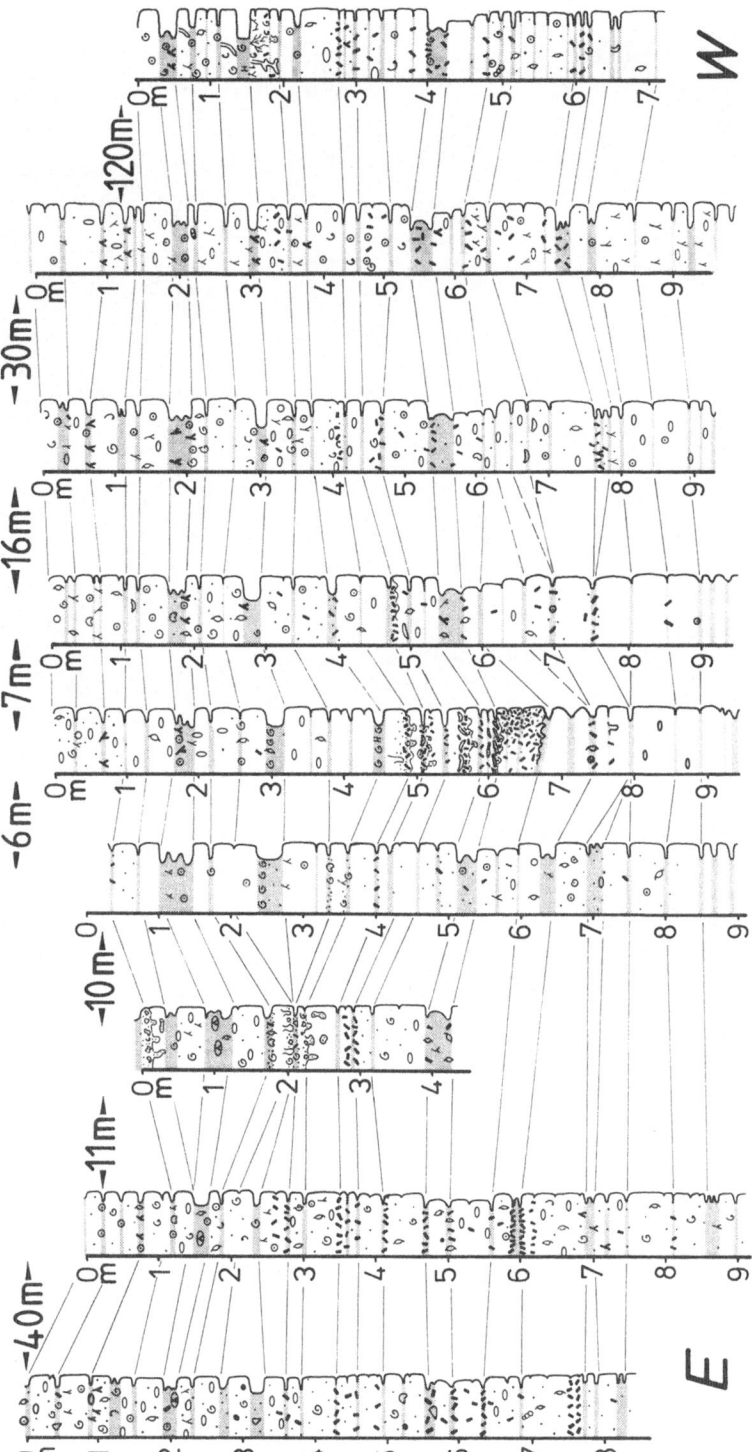

Fig. 87 Bioclastic lags grading laterally into submarine channel fills. Lags occur predominantly in marl beds (shaded). Neuffen Quarry, uppermost Oxfordian, southern Germany. Symbols as in Fig. 91.

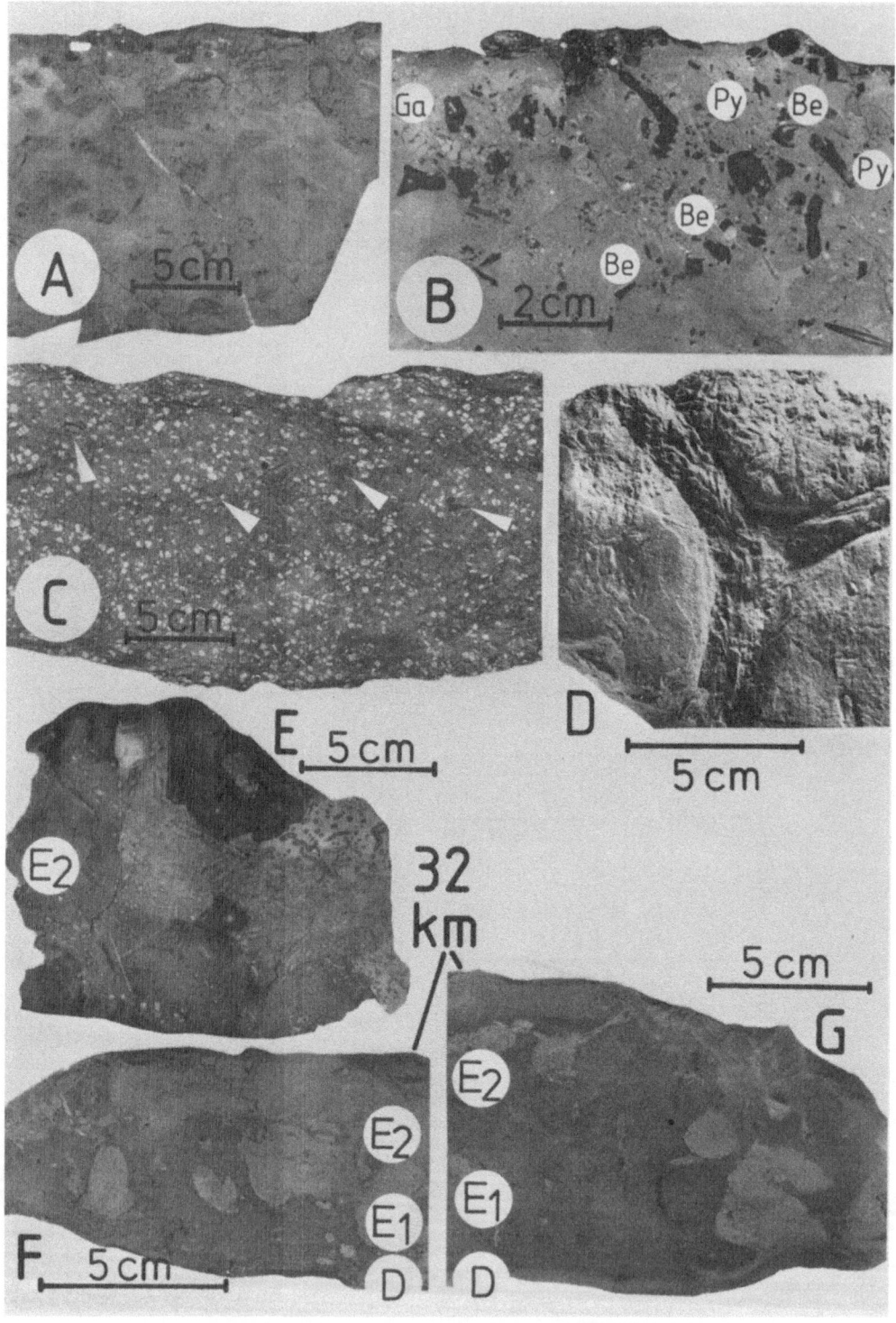

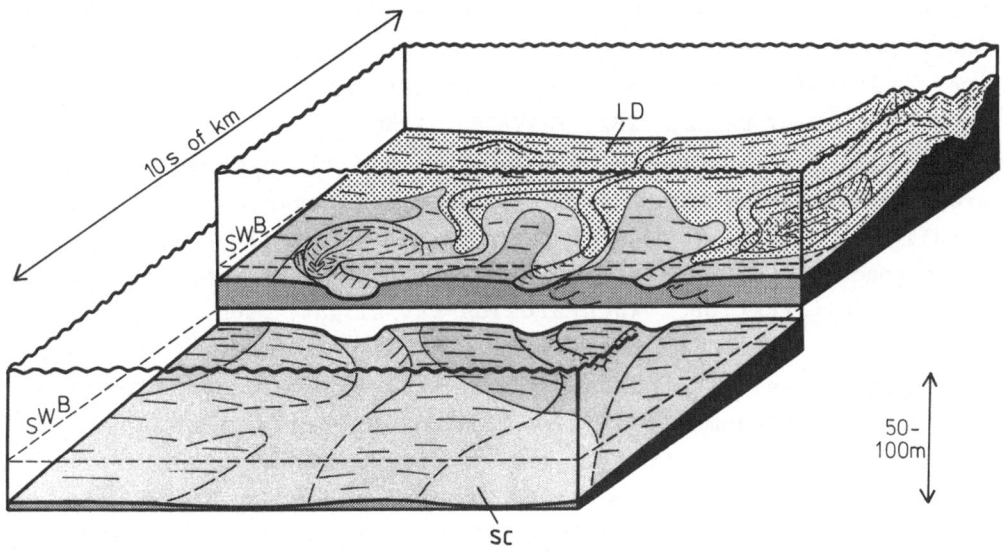

Fig. 89 Sedimentation processes during regressive phases. The storm wave base (SWB) touches bottom, thereby causing lag deposits (LD) on the reef swells and in the basin. Suspension clouds (SC) of fine-grained material are transported in presumably sinusoidal channels below the storm wave base. The eroded sediment is redeposited in submarine fans.

◁**Fig. 88** Petrographic pattern of erosional lags, channel fills, and calcilutite layers in submarine fans. Upper Jurassic, southern Germany.
A: Erosion typically occurs in the marl bed and at the top of the underlaying limestone layer. Detritus is brought into the limestone layer via bioturbation. Neuffen Quarry, Middle Oxfordian.
B: Section from Fig. 88A shows irregular marl intraclasts and fossil remains (e.g., broken belemnites).
C: channel fill consisting of crinoid fragments (often articulated). Arrows indicate sponge fragments. Neuffen Quarry, Upper Oxfordian.
D-G: Graded lutite bed with Bouma-Piper intervals D to E_2 (see Figs. 85, 91 for T_1). The bed belongs to a submarine fan and can be traced for 32km. Post-event bioturbation occurs in two phases (<u>Thalassinoides</u> followed by <u>Chondrites</u>; D,E). Weak grading and thickening of the bed in the channel areas (E); interchannel deposits are characterized by grading, lamination, and microchanneling (F,G). Neuffen Quarry (E,F); Hausen Landslide (D,G). Lowermost Kimmeridgian.

broken (!) belemnite guards, and occasional sponge fragments (Fig. 88C). Most channel fills are equivalent to the so-called "ammonite breccias" which are found throughout the Upper Jurassic in southern Germany (SCHMIDT-KALER, 1962; FREYBERG, 1966).

In the Neuffen Quarry (see section 3.4.1), distinct erosional event layers can be traced laterally from the channels into areas of parallel bedding (Fig. 91). Erosive events within the channels always correspond to either marly lag deposits or "normal" marl beds in the parallel alternation. When erosion was slight, much of the channel lag sediment remained; whereas during intense erosion, little lag material was left in the channels. Thus, the eroded sediment was carried away presumably in the form of suspension clouds. The sediment was redeposited in graded beds composed of calcilutites showing Bouma-Piper intervals, D, E_1, and E_2. These beds are associated with the channels (Fig. 91, T_1 and T_2) and are interpreted to be a portion of a submarine fan. One of these lutite layers (Figs. 88E,F,G; 91, T_1) can be traced in the Neuffen Basin over a distance of 30km (RICKEN, 1985a).

Fig. 90 Lower part of a 35m thick channel system which existed for several million years at the Oxfordian-Kimmeridgian boundary. Limestone layers are shown in black, while marl beds are presented in white. During deposition channel shifting occurs in the lateral and vertical direction. Individual stages of shifting are marked with white circles. Lateral shifting to the NE causes presumably oversteepening of the northeastern channel wall and slumping. Note that erosion occurred predominantly in the marl layers (see Figs. 36E and 91). Neuffen Quarry, southern Germany.

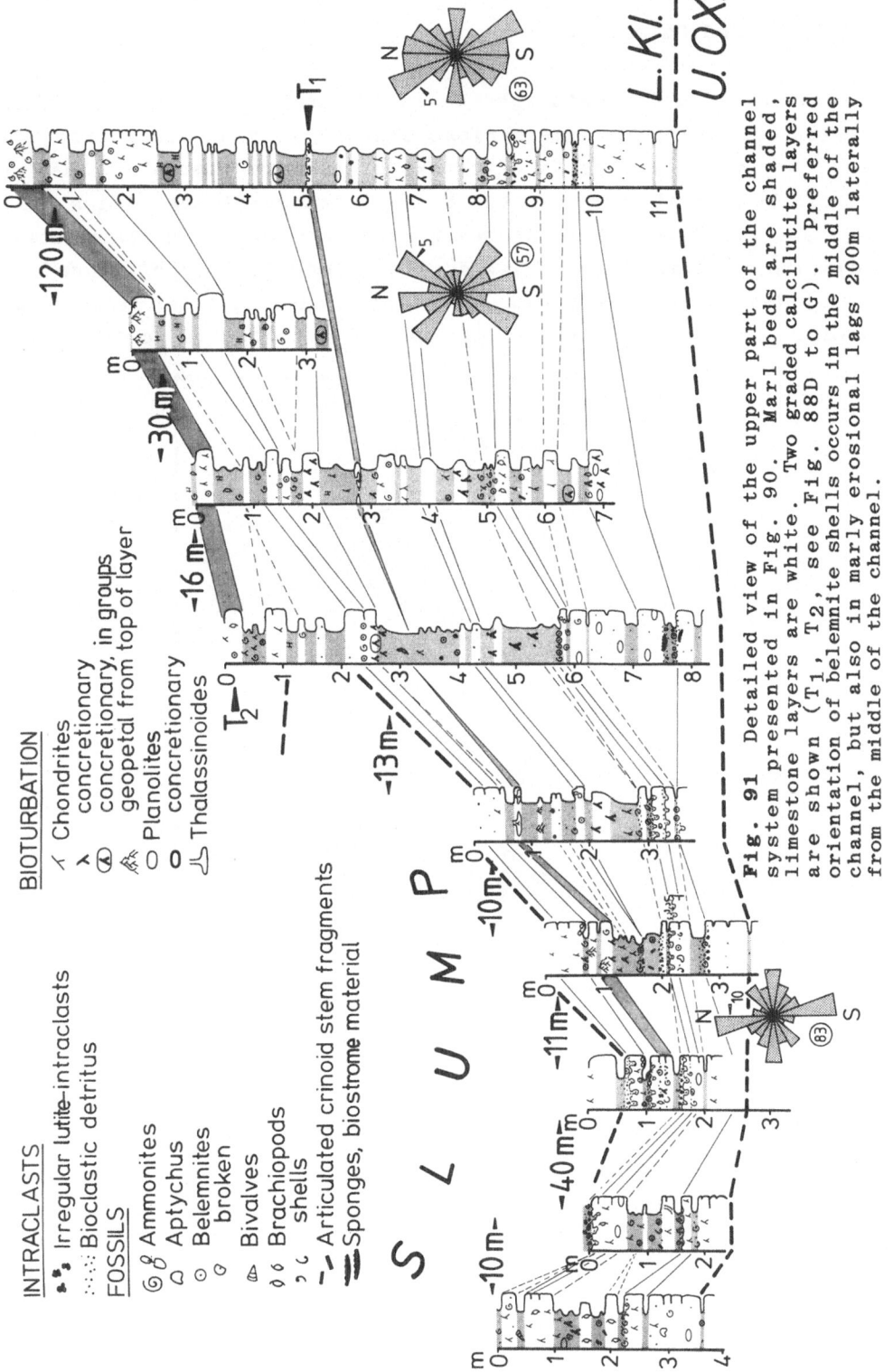

INTRACLASTS

🔸 Irregular lutite-intraclasts

∴ Bioclastic detritus

FOSSILS

🐚 Ammonites
◠ Aptychus
◎ Belemnites broken
🦪 Bivalves
◊ Brachiopods shells
✎ Articulated crinoid stem fragments
▬ Sponges, biostrome material

BIOTURBATION

✓ Chondrites
λ concretionary, in groups
✦ concretionary, in groups geopetal from top of layer
○ Planolites
⊙ concretionary
⊥ Thalassinoides

Fig. 91 Detailed view of the upper part of the channel system presented in Fig. 90. Marl beds are shaded, limestone layers are white. Two graded calcilutite layers are shown (T1, T2, see Fig. 88D to G). Preferred orientation of belemnite shells occurs in the middle of the channel, but also in marly erosional lags 200m laterally from the middle of the channel.

188

8.2.3 Graded Calcilutites

Particularly in the Geisingen Basin and at the western slope of the
Lochen Reef Swell, the alternations contain graded calcilutite beds
which are not clearly associated with submarine channels. They are
especially recognizable near the base of transgressive cycles (Fig.
92), and are interpreted as being somewhere in the spectrum between
turbidite and tempestite. They consist predominantly of graded and
partly laminated lutite and silt-sized carbonates (reef-derived) and
of minor amounts of clastic silicates. The turbidite to tempestite

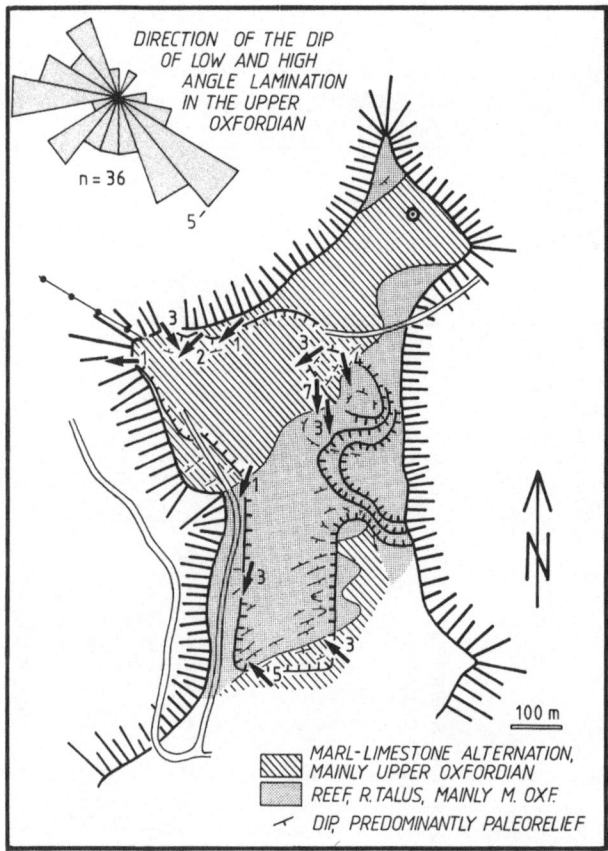

Fig. 92 Upper Oxfordian marl-limestone alternation fills
the interreef areas (hatched) of predominantly Middle
Oxfordian algae-sponge biostromes (shaded). During
deposition of the alternation, the differences in relief
between reef and interreef areas became successively
diminished. Arrows: direction of low and high angle
lamination in some calcilutites in the Oxfordian
marl-limestone alternation. Western slope of the Lochen
Swell, Plettenberg Quarry, southern Germany.

beds represent BOUMA-PIPER intervals B to E_3. Proximal types consist of equal intervals where rippled calcareous silt is overlain by graded and laminated calcilutite (Fig. 93A). Distal types, however, are composed entirely of graded calcilutite which often lack significant lamination. The deposition of those beds from suspension clouds can be supposed from post-event bioturbation (Fig. 93D). Several event beds usually form a single limestone layer (Fig. 93A,B,C). The original $CaCO_3$ content was largely redistributed (see Fig. 53). Sites of preferential diagenetic carbonate dissolution were the base and the top of the turbidite to tempestite beds.

In the Middle and Upper Oxfordian, relic turbidite to tempestite structures are fairly frequent (see Fig. 85), thus the original number of graded layers was much greater than it is now. The dominance of calcilutite deposition during the events was presumably not caused by distal sedimentation modes but instead resulted from a lack of coarse-grained sediment because algae and sponges were the main reef builders. After deposition, the delicate sedimentary structures in lutite event beds (PIPER, 1978; ARTHUR & KELTS, 1981) were easily destroyed even when there was little bioturbation (Fig. 93E,F,G). Later, diagenetic marl seam formation and oxidation of iron compounds during weathering, largely obliterates the bioturbation structure; thus, most of the limestones now appear to be totally homogeneous. The origin of the graded calcilutite beds can be explained by both erosion during storms and slope instability:

1. During transgressions, the erosion facies (that is, the lag deposits; section 8.2.1) presumably shifted to the margins of the basin, where suspension clouds could be generated during storms (DOTT, 1983; WALKER, 1985; AIGNER, 1985). Similar to the processes found in the regressive phases, suspension clouds were transported to the deeper parts of the basins below the storm wave base. Their settling caused graded lutite beds which were not overprinted in their upper parts by wave action.

2. Suspension clouds formed by slumping and during the formation of debris flows, events which frequently occurred in the Upper Jurassic reef-tali of southern Germany. It has already been shown by GWINNER (1962) and T. BRACHERT (Erlangen, personal communication) that slump deposits in the reef talus are often overlain by graded bioclastic beds. Those beds can be regarded as the proximal types of some of the graded calcilutites in the basins.

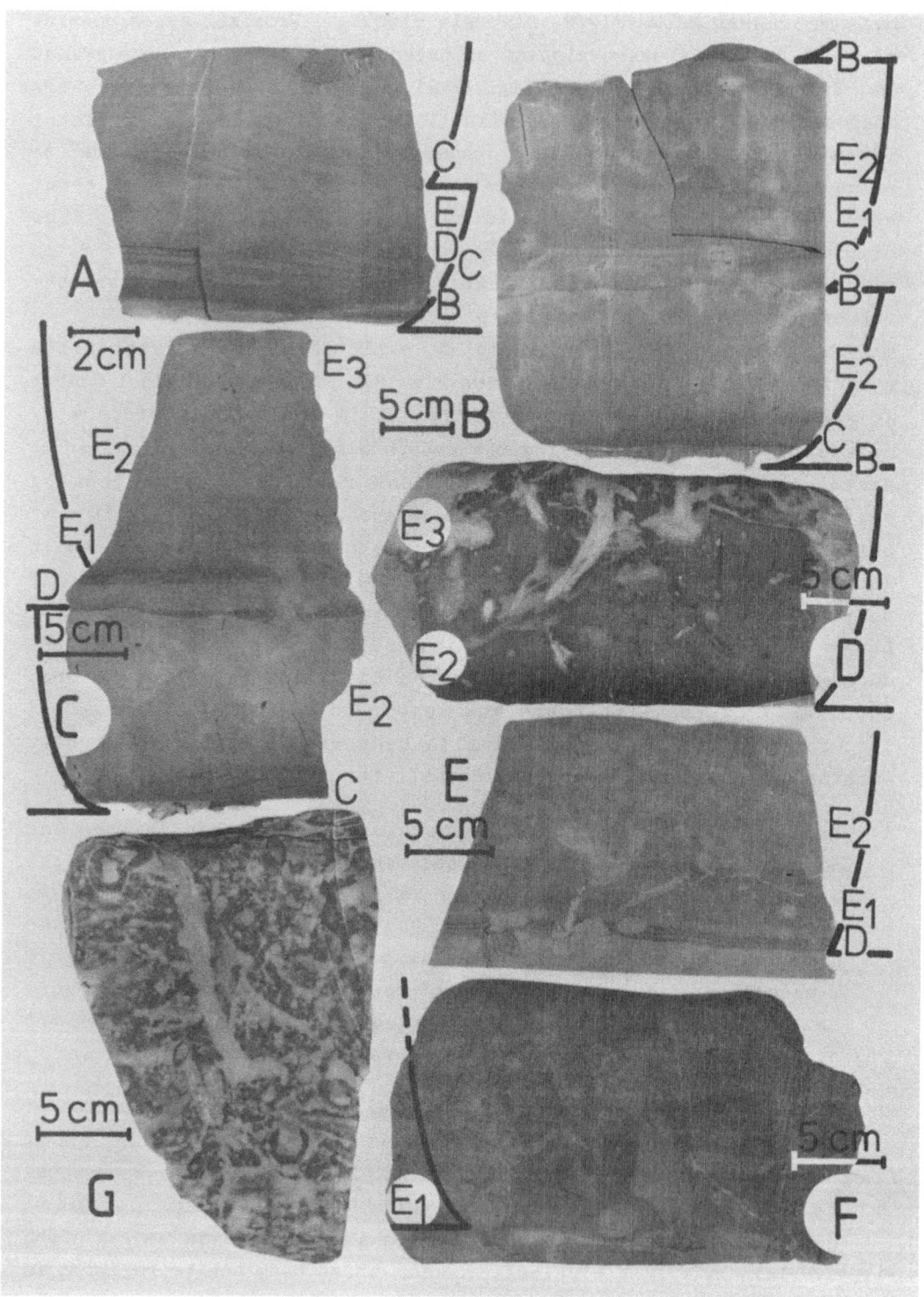

8.3 Conclusions

The rhythmic marl-limestone alternations of the Middle and Upper Oxfordian in southern Germany overprint a rather complex sequence composed of two transgression-regression cycles containing lag deposits, channel and fan systems and graded calcilutite beds. However, primary sedimentary structures were largely destroyed due to bioturbation and diagenesis. The average thickness of the limestone layers remains nearly the same, because diagenetic conditions did not change much through time. However, lateral variations in the limestone layers are related to facies changes. Both subbasins of the Swabian Alb show their own bedding rhythm.

Fig. 93 Sedimentary structures of fine-grained, calcareous turbidites to tempestites which are contained in Upper Jurassic marl-limestone alternations in southern Germany.
A-D: Graded calcilutites with increasing distance from the source. Letters indicate Bouma-Piper intervals (B to E_3). Note, that limestone layers often contain several event beds (see Fig. 53a).
E-G: Progradational destruction of graded calcilutites due to increasing post-event bioturbation (E,F). Common bioturbation pattern in an unweathered limestone layer in the transgressive sequence (G). Geisingen Quarry, Upper Oxfordian (A,B,C); Plettenberg Quarry, Upper Oxfordian (E); Schlatt Landslide, Middle Oxfordian (F).

9 CONCLUSIONS: DIAGENETIC BEDDING

Diagenetic carbonate oscillations were formed during the lithification of calcareous ooze at an overburden of several 100 m of sediment. Alternating zones of $CaCO_3$ dissolution and cementation developed parallel to the primary bedding, which produced a <u>diagenetic stratification</u> in a closed carbonate system. The following major processes occurred:

- The amplitudes of the primary carbonate variations were considerably enhanced.

- Extremely rhythmic alternations originated containing a smaller number of carbonate oscillations as compared to the primary sediment. Diagenetically enhanced carbonate curves tend to have maxima with an angular to convex shape and with narrow, sharp minima.

- Minor elements, especially Mg, were considerably enriched within the relic carbonate of the dissolution-affected marl layers.

The following methods were used to quantify these processes:

- The underlying principle is to measure rock compaction and porosity along with chemical data; e.g., carbonate content. These parameters are mathematically related in the <u>carbonate compaction law</u>, which provides the numerical explanation of diagenetic carbonate curves and enables mass balance calculations to be performed. Compaction was evaluated by using the degree of deformation in originally cylindrical bioturbation tubes.

- Compositional differences in the original sediment and the amount of dissolved and cemented carbonate were computed by performing carbonate mass balance calculations. They are based on the numerical decompaction of the existing rock.

- The concentration of minor elements in the different types of carbonate was evaluated from the bulk minor element concentration contained in the total carbonate fraction and the degree of compaction by using a minor element balance calculation. The enrichment of minor elements in the relic carbonate of the marl beds was explained through the use of the principle of enrichment due to compaction.

Diagenetic marl-limestone alternations were generally formed in bioturbated calcareous oozes with small to moderate variations in carbonate content. Taking into account uncertainties inherent in the calculations, the mean primary carbonate differences between that which are later dissolution and cementation zones (see Fig. 14) were (after bioturbation) between 2.3 and 17.3% with a mean of 5.8% $CaCO_3$. After a phase of mechanical compaction (15 to 55%), cementation of the limestone-layer sediment set in. Carbonate precipitation in the form of pressure shadow structures suggests that differences in the lithostatic stress arose at the grain contacts between the slightly cemented limestone layers and the marl beds. This indicates a self-perpetuating $CaCO_3$ dissolution-reprecipitation process.

It is evident from the simulation models presented here that, in spite of the early onset of diagenetic carbonate redistribution (when porosities were between 50 and 60%), larger diagenetic carbonate oscillations were not generated before the pore space was reduced to less than 30%. After the lithification process came to an end, the amplitude of the carbonate oscillations were 1.5 to 10.3 times higher than those of the original oscillations (mean diagenetic enhancement is by a factor of 5.1). The carbonate redistribution occurred in a predominantly closed system because numerical decompaction of the various sections displays primary porosities which are on the same order of magnitude as those found in recent, fine-grained carbonates.

Chemical compaction caused the enrichment of Mg and in some sections also of Fe within the relic carbonate of the marl beds augmenting their contents by maximum factors of 6 and 2, respectively. On the other hand, the amount of minor elements which were dissolved from the carbonate in the marl beds were completely reprecipitated as minor constituents of the cement in the limestone layers.

The cement content in the limestone layers (about 30 to 45% of the carbonate fraction), the shapes of the carbonate curves, and the degree of enhancement of the primary bedding rhythm were controlled by the amount of compaction at the onset of lithification. If the onset of lithification occurred early in the sediment's diagenetic history, the reduction of the original pore volume was low, and therefore, the pore space of the limestone layer received significant amounts of cement which had to be derived from the carbonate dissolution in the adjacent marl beds. This in turn produced a considerable degree of chemical compaction in the marl beds and changed the carbonate curves of the limestone layers so that they become more angular-shaped as compared to that in the primary sediment. However, the more the thicknesses of the marl beds were reduced due to intense compaction, the more the weakly compacted limestone layers dominate the section and the more the alternations appear to be rhythmic. Simulation models and facies analyses show that this diagenetic rhythmicity was generated independent of whether or not original cyclic stratification was present in the primary sediment (Fig. 94). The primary

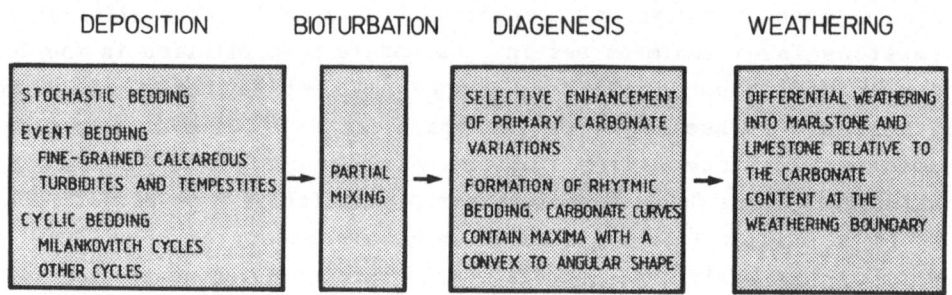

Fig. 94 Summary of the factors which produce lithified marl-limestone alternations.

stratification pattern found in the various types of marl-limestone alternations represents a spectrum between noncyclic and cyclic bedding. Although studied solely in one example (the south German Upper Jurassic), these are in principle stochastic bedding (where only the "noise" of complex systems was preserved), event bedding (e.g., fine-grained turbidites and tempestites), and cyclic bedding (e.g., oceanic production cycles, see DINER & EICHER, 1985; and Milankovitch cycles). It is evident from the detailed carbonate curves constructed that, as compared to the primary sediment, diagenetic bedding produced not more, but a fewer number of carbonate oscillations. Three types of lithified marl-limestone alternations were distinguished, depending on the degree of mechanical compaction which was in turn related to the carbonate content of the primary sediment. The higher the primary carbonate content the lower the degree of mechanical compaction, and usually the earlier the beginning of redistribution process.

Type I   When the primary carbonate content ranged from 60 to 80%, the mechanical compaction of the primary pore space was high (30 to 55%). Therefore, diagenetic carbonate redistribution was low and marl-limestone alternations formed only moderate rhythmicity with sinusoidal carbonate curves and varying maximum carbonate contents. Weathering of the outcrop further weakened the $CaCO_3$ rhythm because smaller carbonate maxima weathered completely to marl.

Type II   Rhythmic marl-limestone alternations originated from sediments with primary carbonate contents of 75 to 90% and reductions in the original volume by 25 to 35% via mechanical compaction. The carbonate curves in the limestone layers display convex shapes and the thickness of the marl beds is approximately equivalent to one half of the thickness of the limestone layers.

Type III When the original carbonate content was high (more than 85%), and therefore, mechanical compaction was low (less than 25%) highly rhythmic, brick-like alternations formed, showing thin, flasery marl joints and angular shapes of the carbonate curves in the limestone layer. The maximum carbonate content (in the middle of the limestone layers) is nearly constant.

10 R E F E R E N C E S

ABU-MAARUF, M. (1975): Feingliederung und Korrelation der Mergelkalkfazies des Untercampans von Misburg, Höver und Wunstorf im ostniedersächsischen Becken.- Berichte Naturhistorische Gesellschaft Hannover 119, 127-204.

AIGNER, T. (1982): Calcareous Tempestites: Storm-dominated stratification in Upper Muschelkalk limestones.- in: EINSELE, G. & SEILACHER, A. (eds.): Cyclic and event stratification, 180-198 (Springer, Berlin).

-- (1985): Storm depositional systems - dynamic stratigraphy in modern and ancient shallow marine sequences.- Lecture Notes in Earth Science 3, 174 p. (Springer, Berlin).

ALVAREZ, L.W.; ALVAREZ, W.; ASARO, F. & MICHEL, H.V. (1980): Extraterrestrial cause for the Cretaceous-Tertiary extinction.- Science 208 (No. 4448), 1095-1108.

ALVAREZ, W.; ENGELDER, T. & GEISER, P.A. (1978): Classification of solution cleavage in pelagic limestones.- Geology 6, 263-266.

-- & LOWRIE, W. (1981): Upper Cretaceous to Eocene pelagic limestones of the Scaglia Rossa are not Miocene turbidites.- Nature 294, 246-294.

--; KAUFFMAN, E.G.; SURLYK, F.; ALVAREZ, L.W.; ASARO, F. & MICHEL, H.V. (1984): Impact theory of mass extinctions and the invertebrate fossil record.- Science 223, 1135-1141.

ARTHUR, M.A. (1976): Sedimentology of Gubbio sequence and its bearing on paleomagnetism.- Memorie Societa Geologica Italiana 15, 9-20.

-- (1979): Sedimentology and geochemical studies of Cretaceous and Paleogene sedimentary rocks: The Gubbio sequence, Part I. Thesis, Princeton Univ., 193 p.

-- ; DEAN, W.E.; BOTTJER, D. & SCHOLLE, P.A. (1984): Rhythmic bedding in Mesozoic-Cenozoic pelagic carbonate sequences: The primary and diagenetic origin of Milancovitch-like cycles.- in: BERGER, A.L. et al. (eds.): Milancovitch and climate, Part 1, 191-222.

-- & FISCHER, A.G. (1977): Upper Cretaceous-Paleocene magnetic stratigraphy at Gubbio, Italy. I Lithostratigraphy and sedimentology.- Geol. Soc. Am. Bull. 88, 367-371.

BALDWIN, B. (1971): Ways of deciphering compacted sediments.- Jour. Sed. Petrol. 41, 293-301.

-- & BUTLER, C.O. (1985): Compaction curves.- Am. Assoc. Petroleum Bull. 69, 622-262.

BAKER, P.A.; GIESKES, J .M. & ELDERFIELD, H. (1982): Diagenesis of carbonates in deep-sea sediments - evidence from Sr/Ca ratios and interstitial dissolved Sr data.- Jour. Sed. Petrol. 52, 71-82.

-- ; KASTNER, M.; BEYERLEE, J.D. & LOCKNER, D.A. (1980): Pressure solution and hydrothermal recrystallization of carbonate sediments - an experimental study.- Marine Geology 38, 185-203.

BANDY, O.L. (1975): Messinian evaporite deposition and the Miocene/ Pliocene boundary, Pasquasia-Capodarso sections, Sicily.- in: SAITO, T. & BURCKLE, L.H. (eds.): Late Neogene epoch boundaries, 49-63 (Micropaleontology Press, New York).

BARKER, P.F.; CARLSON, R.L. et al. (1984): Site 516: Rio Grande Rise.- Init. Repts. DSDP 72, 155-338.

BARRET, P.J. (1964): Residual seams and cementation in Oligocene shell calcarenites, Te Kuiti Group.- Jour. Sed. Petrol. 34, 524-531.

BARRON, E.J.; ARTHUR, M.A. & KAUFFMAN, E.G. (1985): Cretaceous rhythmic bedding sequences: a plausible link between orbital variations and climate.- Earth and Planetary Sci. Lett. 72, 327-340.

BATHURST, R.G.C. (1976): Carbonate sediments and their diagenesis.-
    Develop. Sedimentology 12, 658 p. (Elsevier, Amsterdam).
-- (1980a): Lithification of carbonate sediments.- Sci. Prog., Oxf.
    66, 451-471.
-- (1980b): Deep crustal diagenesis in limestones.- Revista Inst.
    Investigaciones Geol., Barcelona 34, 89-100.
-- (1983): Neomorphic spar versus cement in some Jurassic grainstones:
    significance for evaluation of porosity evolution and compaction.-
    Jour. Geol. Soc. London 140, 229-237.
-- (1984): Origin of bedding planes defined by pressure-dissolution
    seams in Phanerozoic platform limestones (abstr.).- 7th Meeting
    Carbon. Sedimentologists, Liverpool.
BAUSCH, W.M. (1968): Clay content and calcite crystal size of lime-
    stones.- Sedimentology 10, 71-75.
-- (1982): Tonmineralprovinzen im Malmkalken.- Erlanger Forschungen
    B8, 78 p.
-- ; FATSCHEL, J. & HOFMANN, D. (1982): Observations on well-bedded
    Upper Jurassic limestones.- in: EINSELE, G. & SEILACHER, A. (eds.):
    Cyclic and event stratification, 54-62 (Springer, Berlin).
-- & POLL, K.G. (1984): Das Profil Loser/Altaussee - Bindeglied
    zwischen alpinem und süddeutschem Malm.- Geol. Rundschau 73,
    351-369.
BEACH, D.K. & SCHUMACHER, A.L. (1982): Stanley field, North Dakota,
    economic and quantitative significance of mechanically compacted
    shallow water limestone.- Am. Assoc. Petroleum Bull. 66, 547-548.
BENSON, R.H. (1975): Ostracods and Neogene history.- in: SAITO, T. &
    BURCKLE, L.H. (eds.): Late Neogene epoch boundaries, 41-48
    (Micropaleonotology Press, New York).
BERGER, A.L.; IMBRIE, J.; HAYS, J.; KUKLA, G. & SALTZMAN, B. (1984,
    eds.): Milankovitch and climate. (The Hague, Reidel).
BERGER, W.H.; EKDALE, A.A. & BRYANT, P.P. (1979): Selective preserv-
    ation of burrows in deep-sea carbonates.- Marine Geology 32,
    204-230.
-- & RAD, U.V. (1972): Cretaceous and Cenozoic sediments from the
    Atlantic Ocean.- Init. Rep. Deep Sea Drilling Project 14, 787-954.
BERNER, R.A. (1980): Early diagenesis, 241 p. (University Press,
    Princeton).
-- (1984): Sedimentary pyrite formation: an update.- Geochim.
    Cosmochim. Acta 48, 605-615.
BHATTACHARYYA, A. & FRIEDMAN, G.M. (1983): Mineralogical and
    paramorphic textural changes generated in modern ooids by heat and
    compaction.- Geology, 11, 596-598.
-- (1984): Experimental compaction of ooids under deep-burial
    diagenetic temperatures and pressures.- Jour. Sed. Petrol. 54,
    362-372.
BIJU-DUVAL, B. & MONTADERT, L. (1976, eds.): Structural history of the
    Mediterranean basin, 448 p. (Editions Technique, Paris).
BOLES, J.R. & FRANKS, S.G. (1979): Clay diagenesis in Wilcox
    sandstones of southwest Texas: Implications of smectite diagenesis
    on sandstone cementation.- Jour. Sed. Petrol. 49, 55-70.
BORTOLOTTI, V.; PASSERINI, P.; SAGRI, M. & SESTINI, G. (1970): The
    miogeosynclinal sequences.- Sedimentary Geol. 4, 341-444.
BOTTJER, D.J.; ARTHUR, M.A. & DEAN, W.E. (1985): Rhythmic bedding in
    Upper Cretaceous chalks.- Cycles and periodicity in geologic
    events, evolution and stratigraphy, Research Symp. Princeton, p. 4
    (abstract).
BOUMA, A.H. (1962): Sedimentology of some flysch deposits, 168 p.
    (Elsevier, Amsterdam).
BRAND, U. & VEIZER, J. (1980): Chemical diagenesis of a multicomponent
    carbonate system - I. Trace elements.- Jour. Sed. Petrol. 50,
    1219-1236.

BROLSMA, M.J. (1976): Discussion of the arguments concerning the paleo-environmental interpretation of the arenazzolo in Capo Rosello and Eraclea Minora (s. Sicily, Italy).- Mem. Soc. Geol. Italiana 16, 153-157.

BROMLEY, R.G. & EKDALE, A.A. (1984): Chondrites: a trace fossil indicator of anoxia in sediments.- Science 224, 872-874.

BURGER, G. (1982): Tonmineralogische und sedimentpetrographische Untersuchungen in der untersten Kreide des östlichen Helvetikums.- Schweiz. mineral. petrogr. Mitt. 62, 369-414.

BUSNARDO, R. (1963): Le stratotype du Barrémien. I. Lithologie et macrofaune.- Coll. Crétacé inf. Lyon, Mém. B.R.G.M. Paris 34 (1965), 99-128.

BUXTON, T.J. & SIBLEY, D.F. (1981): Pressure solution features in a shallow buried limestoue. Jour. Sed. Petrol. 51, 19-26.

CAMPOS, H.S. & HALLAM, A. (1979): Diagenesis of English Lower Jurassic limestones as inferred from oxygen and carbon isotope analysis.- Earth and Planet. Sci. Lett. 45, 23-31.

CATALANO, R,; RUGGIERI, G. & SPROVIERI, R. (1975, eds.): Messinian evaporites in the Mediterranean.- Memorie Società Geologica Italiana 14 (1976), 385 p.

CHANDA, S.K.; BHATTACHARYYA, A. & SARKAR, S. (1977): Deformation of ooids by compaction in the Precambrian Bhander Limestone, India: implications for lithification.- Geol. Soc. Am. Bull. 88, 1577-1585.

CITA, M.B. (1972): Il significato della transgressione Pliocenica alla luce delle nuove scoperte nel Mediterraneo.- Riv. Ital. Paleont. 78, 527-594.

-- (1973): Mediterranean evaporite: Paleontological arguments for a deep-basin desiccation model.- in: DROOGER, C.W. (ed.): Messinian events in the Mediterranean, 206-223 (North Holland Publishing Comp., Amsterdam).

COOGAN, A.H. (1970): Measurements of compaction in oolitic grainstones.- Jour. Sed. Petrol. 40, 921-939.

CORRENS, C.W. (1949): Growth and dissolution of crystals under linear pressure.- Discussions Faraday Soc. 5, 267-271.

COTILLON, P. (1971): Le Crétacé inférieur de l'arc subalpin de Castellane entre l'Asse et le Var. Stratigraphie et sédimentologie.- Mémoires du B.R.G.M. 68, 313 p.

-- ; FERRY, S.; GAILLARD, C.; JAUTEE, E.; LATREILLE, G. & RIO, M. (1980): Fluctuation de paramètres du milieu marin dans le domaine vocontien (France SE) au Crétacé inférieur: mise en évidence par l'étude des formations marno-calcaires alternates.- Bull. Soc. Géol. France 22, 735-744.

-- & RIO (1984): Cyclic sedimentation in the Cretaceous of Deep Sea Drilling Project sites 535 and 540 (Gulf of Mexico), 534 (Central Atlantic), and in the Vocontian Basin (France).- in: BUFFLER, R.T.; SCHLAGER, W. et al. (eds.): Init. Repts. DSDP 77. 339-376.

CRIMES, T.P. (1975): The stratigraphical significance of trace fossils.- in: FREY, R.W. (ed.): The study of trace fossils, 109-130 (Springer, New York).

CZERNIAKOWSKI, L.A.; LOHMANN, K.C. & WILSON, J.L. (1984): Closed-system marine burial diagenesis: isotopic data from the Austin Chalk and its components.- Sedimentology 31, 863-877.

DARMEDRU. C. (1984): Variations au taux de sédimentation et oscillations climatiques lors du dépôts des alternances marne-calcaire pélagiques. Exemple du Valanginien supérieur - Vocontien (SE de la France).- Bull. Soc. Géol. France 26, 63-70.

--; COTILLON, P. & RIO, M. (1982): Rhythmes climatiques et biologiques en milieu marin pélagique. Leurs relations dans les dépôts crétacés alternants du bassin vocontien (SE France).- Bull. Soc. Geol. France 7, 627-640.

DEAN, W.E. & GRADNER, J.V. (1985): Milankovitch cycles in Neogene deep-sea sediments.- Cycles and periodicity in geologic events, evolution and stratigraphy, Research Symp. Princeton, p. 7 (abstract).

DE BOER, P.L. & WONDERS, A.A.H. (1984): Astronomically induced rhythmic bedding in Cretaceous pelagic sediments near Moria (Italy).- in: BERGER, A.; IMBRIE, J. et al. (eds): Milankovitch and climate, 177-190 (The Hague, Reidel).

DOTT, R.H. (1983): Episodic sedimentation - how normal is average? How rare is rare? Does it matter?- Jour. Sed. Petrol. 53, 5-23.

DROOGER, C.W. (1973, ed.): Messinian events in the Mediterranean, 272 p. (North-Holland Publishing Comp., Amsterdam).

EDER, F.W. (1971): Riff-nahe detritische Kalke bei Balve im Rheinischen Schiefergebirge.- Göttinger Arb. Geol. Paläontol. 10, 66 p.

-- (1982): Diagenetic redistribution of carbonate, a process in forming limestone-marl alternations (Devonian and Carboniferous, Rheinisches Schiefergebirge, W. Germany).- in: EINSELE, G. & SEILACHER, A. (eds.): Cyclic and event stratification, 98-112 (Springer, Berlin).

EICHER, D.L. & DINER, R. (1985): Foraminifera as indicators of water mass in the Cretaceous Greenhorn sea, Western Interior.- Soc. Econ. Paleont. Mineral. field trip guidebook 4, 60-71.

EINSELE, G. (1977): Range, velocity and material flux of compaction flow in growing sedimentary sequences.- Sedimentology 24, 639-655.

-- (1982): Limestone-marl cycles (Periodites): Diagnosis, significance, causes - a review.- in: EINSELE, G. & SEILACHER, A. (eds.): Cyclic and event stratification, 8-53 (Springer, Berlin).

-- (1985): Response of sediments to sea-level changes in differing subsiding storm-dominated marginal and epeiric basins.- in: BAYER, U. & SEILACHER, A. (eds.): Sedimentary and evolutionary cycles.- Lecture Notes in Earth Sciences 1, 127-162.

-- & MOSEBACH, R. (1955): Zur Petrographie, Fossilerhaltung und Entstehung der Gesteine des Posidonienschiefers im Schwäbischen Jura.- Neues Jahrb. Geol.   Paläontol. Abh. 101, 319-430.

-- & SEILACHER, A. (1982, eds.): Cyclic and event stratification, 536 p. (Springer, Berlin).

EKDALE, A.A. & BROMLEY, R.G. (1984): Sedimentology and ichnology of the Cretaceous-Tertiary boundary in Denmark: implications for the causes of the terminal Cretaceous extinction.- Jour. Sed. Petrol. 54, 681-703.

--; MULLER, L.N. & NOVAK, M.T. (1984): Quantitative ichnology of modern pelagic deposits in the abyssal Atlantic.- Paleogeogr., Paleoclim., Paleoec. 45, 189-223.

ELDER, W.P. & KIRKLAND, J.I. (1985): Stratigraphy and depositional environments of the Bridge Creek limestone member of the Greenhorn Limestone at Rock Canyon Anticline near Pueblo, Colorado.- Soc. Econ. Paleont. Mineral. field trip guidebook 4, 122-134.

ELDERFIELD, H. & GIESKES, J.M. (1982): Sr isotopes in interstitial waters of marine sediments from Deep Sea Drilling Project cores.- Nature 300, 493-496.

ENGELDER, T.; GEISER, P.A. & ALVAREZ, K. (1981): Role of pressure solution and dissolution in geology.- Geology 9, 44-45.

ENGELHARD, W. (1977): The origin of sediments and sedimentary rocks. 359 p. (E. Schweizerbart, Stuttgart).

ERNST, G.; SCHMID, F. & KLISCHIES, G. (1979): Multistratigraphische Untersuchungen in der Oberkreide des Raumes Braunschweig-Hannover.- in: WIEDMANN, J. (ed.): Aspekte der Kreide Europas.- I.U.G.S. Series A 6, 11-46.

FISCHER, A.G. (1980): Gilbert - bedding rhythms and geochronology.- in: YOCHELSON, E.L. (ed.): The scientific ideas of G.K. Gilbert.- Geol. Soc. Am. Spec. Paper 183, 93-104.

-- & ARTHUR, M.A. (1977): Secular variations in the pelagic realm.-
in: COOK, H.E. & ENOS, P. (eds.): Deep-water carbonate
environments.- Soc. Econ. Paleont. Mineral. Spec. Publ. 25, 19-50.

--; HERBERT, T. & PREMOLI SILVA, J. (1985): Carbonate bedding cycles
in Cretaceous pelagic and hemipelagic sequences.- Soc. Econ.
Paleont. Mineral. field trip guidebook 4, 1-10.

FLÜGEL, H. (1968): Die Lithogenese der Oberalmer Schichten und der
mikritischen Plassen-Kalke (Tithonium, Nördliche Kalkalpen).- Neues
Jahrb. Geol. Paläontol. Abh. 123, 249-280.

FLÜGEL, H.W. (1968): Some notes on the insoluble residues in
limestones.- in: MÜLLER, G. & FRIEDMAN, G.M. (eds.): Recent
developments in carbonate sedimentology in central Europe, 46-54
(Springer, Berlin).

-- & WEDEPOHL, K.H. (1967): Die Verteilung des Strontiums in
oberjurassischen Karbonatgesteinen der nördlichen Kalkalpen.-
Contr. Mineral. Petrol. 14, 229-249.

FOLK, R.L. & LAND, L.S. (1975): Mg/Ca ratio and salinity: two controls
over crystallization of dolomite.- Am. Assoc. Petroleum Geol. Bull.
59, 60-68.

FÜCHTBAUER, H. & MÜLLER, G. (1977): Sedimente und Sedimentgesteine
(Sediment-Petrologie Teil II), 784 p. (Schweizerbart, Stuttgart).

FREYBERG, B.V. (1966): Der Faziesverband im unteren Malm Frankens.-
Erlanger Geol. Abh. 62, 3-92.

GARDNER, J.V.; DEAN, W.G. & JANSA, L. (1977): Sediments recovered from
the northwest African continental margin.- Init. Repts. DSDP 41
1121-1134.

GARRISON, R.E.: (1981): Diagenesis of oceanic carbonate sediments: A
review of the DSDP perspective.- Soc. Econ. Paleont. Mineral. Spec.
Publ. 32, 181-207.

-- & KENNEDY, W.J. (1977): Origin of solution seams and flaser
structures in the Upper Cretaceous chalks of southern England.-
Sedimentary Geology 19, 107-137.

GEBHARD, G. (1983): Stratigraphische Kondensation am Beispiel
mittelkretazischer Vorkommen im perialpinen Raum.- Thesis, Univ.
Tübingen, 145 p.

GEISER, P.A. & SANSONE, S. (1981): Joints, microfractures, and the
formation of solution cleavage in limestones.- Geology 9, 280-285.

GIESKES, J.M. (1981): Deep-sea drilling interstitial water studies:
implications for chemical alteration of the oceanic crust, layers I
and II.- Soc. Econ. Paleont. Mineral. Spec. Publ. 32, 149-167.

-- (1965): Chemistry of interstitial waters of marine sediments.- An.
Rev. Earth and Planet. Sci. 3, 433-453.

GILBERT, G.K. (1895): Sedimentary measurement of geologic time.- Jour.
Geology 3, 121-127.

GLUYAS, J.G. (1984): Early carbonate diagenesis within Phanerozoic
shales and sandstones of the NW European shelf.- Clay Minerals 19,
309-321.

GWINNER, M.P. (1962): Subaquatische Gleitungen und resedimentäre
Breccien im Weißen Jura der Schwäbischen Alb (Württemberg).-
Zeitschr. deutsche geol. Gesell. 113, 571-590.

-- (1976): Origin of Upper Jurassic limestones of the Swabian Alb
(southwest Germany).- Contributions to Sedimentology 5, 75 p.

GYGI, R. (1969): Zur Stratigraphie der Oxford-Stufe der Nordschweiz
und des süddeutschen Grenzgebietes.- Beiträge geol. Karte Schweiz
N.F. 136, 123 p.

HALLAM, A. (1964): Origin of the limestone-shale rhythms in the Blue
Lias of England: A composite theory.- Jour. of Geol. 72, 157-168.

HAMILTON, E.L. (1974): Prediction of deep-sea sediment properties:
state of art.- in: INDERBITZEN, A.L. (ed.): Deep Sea Sediments:
Physical and Mechanical Properties, 1-44 (Plenum Press, New York).

-- (1976): Variations of density and porosity with depth in deep-sea
sediments.- Jour. Sed. Petrol. 46, 280-300.

HÄNTZSCHEL, W. (1975): Trace fossils and problematica.- in: MOORE, R. & TEICHERT, C. (eds.): Treatise on invertebrate paleontology, Part W, 269 p. (Univ. Kansas Print. Serv.).

HARMS, J.C. & CHOQUETTE, P.W. (1965): Geologic evaluation of a gamma-ray porosity device.- in: 6th Ann. SPWLA Logging Symposium, vol. 2, C1-C37.

HATTIN, D.E. (1971): Widespread, synchronously deposited, burrow-mottled limestone beds in Greenhorn Limestone (Upper Cretaceous) of Kansas and southern Colorado.- Am. Assoc. Petrol. Bull. 55, 412-431.

-- (1985): Distribution and significance of widespread, time-parallel pelagic limestone beds in Greenhorn Limestone (Upper Cretaceous) of the Central Great Plains and southern Rocky Mountains.- Soc. Econ. Paleont. Mineral. field trip guidebook 4, 28-37.

HEIMANN, K.O. (1977): Die Fazies des Messins und untersten Pliozäns auf den Ionischen Inseln (Zakynthos, Kephallinia, Korfu/Griechenland) und auf Sizilien.- Thesis, Techn. Univ. München, 158 p.

HENNINGSMOEN, G. (1974): A comment. Origin of limestone nodules in the Lower Paleozoic of the Oslo region.- Norsk Geologisk Tidsskrift 54, 401-412.

HERRMANN, A.G. (1975): Praktikum der Gesteinsanalyse, 204. (Springer, Berlin).

HILLER, K. (1964): Über die Bank- und Schwammfazies des Weißen Jura der Schwäbischen Alb (Württemberg).- Arbeiten Geol. Paläont. Inst. TH Stuttgart NF 40, 190 p.

HINTE, J.E. van (1976a): A Jurassic time scale.- Am. Assoc. Petroleum Bull. 60, 489-497.

-- (1976b): A Cretaceous time scale.- Am. Assoc. Petroleum Bull. 60, 498-516.

HÖLLER, H. & WALITZI, E.M. (1965): Mineralogische Untersuchungen an den Oberalmer Schichten und an den mikritischen Plassenkalken, Nördliche Kalkalpen.- Neues Jahrb. Geol. Paläont. Mh. 123, 552-555.

HSÜ, K. (1983): The Mediterranean was a desert. A voyage of the Glomar Challenger, 197 p. (Princeton Univ. Press).

HUDSON, J.D. (1975): Carbon isotopes and limestone cement.- Geology 3, 19-20.

JANOWSKY, W. (1970): Empirical investigation of some factors affecting elastic wave velocities in carbonate rocks.- Geophys. Prosp. 18, 103-118.

JEANS, C.V. (1980): Early submarine lithification in the Red Chalk and Lower Chalk of eastern England: A bacterial control model and its implications.- Proc. Yorkshire Geol. Soc. 43 (Part 2, No. 6), 81-157.

JØRGENSEN, N.O. (1983): Dolomitization in chalk from the North Sea Central Graben.- Jour. Sed. Petrol. 53, 557-564.

KAHLE, C.F. (1966): Some observations on compaction and consolidation in ancient oolites.- Compass 44, 19-29.

KATZ, A.; SASS, E.; STARINSKY, A. & HOLLAND, H.D. (1972): Strontium behavior in the aragonite-calcite transformation: An experimental study at 40-90°C.- Geochim. Cosmochim. Acta 36, 481-496.

KEITH, M.L. (1982): Violent volcanism, stagnant oceans and some inferences regarding petroleum, strata-bound ores and mass extinctions.- Geochim. Cosmochim. Acta 46, 2621-2637.

KELLER, G.H. & BENNETT, R.H. (1970): Variations in the mass physical properties of selected submarine sediments.- Marine Geology 9, 215-223.

-- ; LAMBERT, D.N. & BENNETT, R.H. (1979): Geotechnical properties of continental slop deposits - Cape Hatteras to Hydrographer Canyon.- Soc. Econ. Paleont. Mineral. Spec. Publ. 27, 131-151.

KENNEDY, W.J. (1975): Trace fossils in carbonate rocks.- in: FREY, R.W. (ed.): The study of trace fossils, 377-397 (Springer, Berlin).

-- & ODIN, G.S. (1982): The Jurassic and Cretaceous time scale in 1981.- in: ODIN, G.S. (ed.): Numerical dating in stratigraphy, 557-592 (John Wiley, Chichester).

KENT, D.V. (1981): Asteroid extinction hypothesis.- Science 214, 648-650.

KERCKHOVE, C. & ROUX, M. (1976): Carte géologique de la France, 1:50000, XXXV-42, Castellane.- Bureau de recherches géologiques et minières, 39 p.

KETTENBRINK, E.C. & MANGER, W.L. (1971): A deformed marine pisolite from the Plattsburg Limestone (Upper Pennsylvanian) of southern Kansas.- Jour. Sed. Petrol. 41, 435-443.

KINSMAN, D.J.J. (1969): Interpretation of Sr^{2+} concentrations in carbonate minerals and rocks.- Jour. Sed. Petrol. 39, 486-508.

KNOBLAUCH, G. (1963): Sedimentpetrographische und geochemische Untersuchungen an Weißjurakalken der geschichteten Fazies im Gebiet von Urach und Neuffen.- Thesis, Univ. Tübingen, 106 p.

KOEPNICK, R.B. (1985): Distribution and permeability of stylolite-bearing horizons within a Lower Cretaceous carbonate reservoir in the Middle East.- 60th An. Tech. Conf. of Soc. Petroleum Engineers, Las Vegas, NV, 7 p.

KÖHLER, K.E. (1971): Zur Sedimentologie der Grenzschichten Dogger/Malm in Südwestwürttemberg.- Arbeiten Geol. Paläontol. Inst. TH Stuttgart NF 64, 90 p.

KOPF, M. (1983): Über die Kreidekalkdiagenese.- Zeitschr. geol. Wiss. 11, 1443-1451.

KRANZ, J.R. (1976): Strontium - ein Fazies-Diagenese-Indikator im oberen Wettersteinkalk (Mittel-Trias) der Ostalpen.- Geol. Rundschau 65, 593-165.

LABUDE, C. (1983): Sedimentologie, Subsidenz und die Verbreitung sandschaliger und planktonischer Foraminiferen im Maastricht des zentralen Apennin Italiens.- Thesis Univ. Tübingen, 95 p.

LIPPMANN, F. (1973): Sedimentary carbonate minerals, 228 p. (Springer Verlag, Berlin).

LOCKRIDGE, I.P. & SCHOLLE, P.A. (1978): Niobrara gas in eastern Colorado and northwestern Kansas.- Rocky Mount. Assoc. Geol. Symp. 1978, 35-49.

LOWRIE, W. & ALVAREZ, W. (1977): Upper Cretaceous-Paleocene magnetic stratigraphy at Gubbio, Italy. III Upper Cretaceous magnetic stratigraphy.- Geol. Soc. Am. Bull. 88, 374-377.

LUTERBACHER, H.P. & PREMOLI SILVA, I. (1962): Note préliminaire sur une revision du profile de Gubbio, Italie.- Riv. Italiana Paleontologia e Stratigrafia 68, 253-288.

MANUS, R.W. & COOGAN, A.H. (1974): Bulk volume reduction and pressure-solution derived cement.- Jour. Sed. Petrol. 44, 466-471.

MARSCHNER, H. (1968): Relationship between carbonate grain size and noncarbonate content in carbonate sedimentary rocks.- in: MÜLLER, G. & FRIEDMAN, G.M. (eds.): Recent developments in carbonate sedimentology in central Europe, 55-57 (Springer, Berlin).

MATTER, A. (1974): Burial diagenesis of pelitic and carbonate deep-sea sediments from the Arabian Sea.- Init. Rep. DSDP 23, 421-470.

MATTES, B.W. & MOUNTJOY, E.W. (1980): Burial dolomitization of the Upper Devonian Miette buildup, Jasper National Park, Alberta.- Soc. Econ. Paleont. Mineral. Spec. Publ. 28, 259-297.

MAYER, L.A. (1980): Deep-sea carbonates: Physical property relationships and the origin of high-frequency acoustic reflectors.- Marine Geology 38, 165-183.

MC LEAN, D.M. (1982): Deccan volcanism and the Cretaceous-Tertiary transition scenario: A unifying causal mechanism.- in: RUSSELL, D.A. & RICE, G. (eds.): Cretaceous-Tertiary extinctions and possible terrestrial and extraterrestrial causes. Nat. Mus. Canada Syllog. 39, 143-144.

MEISCHNER, K.D. (1964): Allodapische Kalke; Turbidite in Riff-nahen Sedimentationsbecken.- Develop. Sedimentology 3, 156-191.

MERINO, E.; ORTOLEVA, P. & STRICKHOLM, P. (1983): Generation of evenly-spaced pressure-solution seams during (late) diagenesis: a kinetic theory.- Contrib. Mineral. Petrol. 82, 360-370.

MEYERS, W.J. & HILL B.E. (1983): Quantitative studies of compaction in Mississippian skeletal limestones, New Mexico.- Jour. Sed. Petrol. 53, 321-242.

MICHARD, G. (1971): Contribution à l'étude de l'entraînement des éléments traces dans le calcite lors de sa précipitation.- Chem. Geol. 8, 311-327.

MILANKOVITCH, M. (1930): Mathematische Klimalehre und astronomische Theorie der Klimaschwankungen.- in: KÖPPEN, W. & GEIGER, R. (eds.): Handbuch der Klimatologie, vol. 1, 176 p. (Borntraeger, Berlin).

MIMRAN, Y. (1977): Chalk deformation and large-scale migration of calcium carbonate.- Sedimentology 24, 333-360.

MITRA, S. & BEARD, W.D. (1980): Theoretical models of porosity reduction by pressure solution for well-sorted sandstones.- Jour. Sed. Petrol. 50, 1347-1360.

MORROW, D.W. & MAYERS, J.R. (1977): Simulation of limestone diagenesis - a model based on strontium depletion.- Canadian Jour. Earth Sci. 15, 376-396.

MUCCI, A. & MORSE, J.W. (1983): The incorporation of Mg^{2+} and Sr^{2+} into calcite overgrowths: influences of growth rate and solution composition.- Geochim. Cosmochim. Acta 47, 217-233.

NEUGEBAUER, J. (1973): The diagenetic problem of chalk.- Neues Jahrb. Geol. Paläontol. Abh. 143, 136-156.

-- (1974): Some aspects of cementation in chalk.- in: HSÜ, K.J. & JENKYNS, C. (eds.): Pelagic sediments: on land and under the sea.- Int. Assoc. Sedim. Spec. Publ. 1, 149-176.

O'KEEFE, J.D. & AHRENS, T.J. (1982): Impact mechanics of the Cretaceous-Tertiary extinction bolide.- Nature 298, 123-127.

OSMOND, J.K. (1981): Quarternary deep-sea sediments: Accumulation rates and geochronolgy.- in: Emiliani, C. (ed.): The oceanic lithosphere, 1329-1371.

PLESSMANN, W. (1964): Gesteinslösung, ein Hauptfaktor beim Schieferungsprozess.- Geol. Mitt. 4, 69-82.

-- (1966): Diagenetische und kompressive Verformung in der Oberkreide des Harz-Nordrandes sowie im Flysch von San Remo.- Neues Jahrb. Geol. Paläontol. Mh. 8, 480-493.

PERRIER, R. & QUIBLIER, J. (1974): Thickness changes in sedimentary layers during compaction history; methods for quantitative evaluation.- Am. Assoc. Petroleum Geol. Bull. 38, 507-520.

PINGITORE, N.E. (1976): Vadose and phreatic diagenesis: Processes, products, and their recognition in corals.- Jour. Sed. Petrol. 46, 985-1006.

-- (1982): The rule of diffusion during carbonate diagenesis.- Jour. Sed. Petrol. 52, 27-39.

PIPER, D.J.W. (1978): Turbidite muds and silts on deep-sea fans and abyssal plains.- in: STANLEY, D.J. & KELLING, G. (eds.): Sedimentation in submarine canyons, fans and trenches, 163-176 (Hutchinson & Ross, Stoudsburg).

POLLASTRO, R.M. & MARTINEZ, G.J. (1985): Whole-rock, insoluble residue, and clay mineralogie of marl, chalk, and bentonite; Smoky Hill shale member, Niobrara Formation near Pueblo, Colorado - depositional and diagenetic implications.- Soc. Econ. Paleont. Mineral. field trip guidebook 4, 215-222.

PRAY, L.C. (1960): Compaction in calcilutites.- Geol. Soc. Am. Bull. 71, p. 1946 (abstract).

-- (1966): Informal comments on calcium carbonate cementation. Soc. Econ. Paleont. Mineral. Techn. Session on Lithification and Diagenesis, St. Louis Meetings.

PREMOLI SILVA, I. (1977): Upper Cretaceous-Paleocene stratigraphy at Gubbio, Italy. II Biostratigraphy.- Geol. Soc. Am. Bull. 88, 371-374.

RAMPINO, M.R. (1982): A non-catastrophist explanation for the iridium anomaly at the Cretaceous-Tertiary boundary.- Geol. Soc. Am. Spec. Paper 190, 455-460.

RAMSAY, J.G. & HUBER, M.I. (1983): The techniques of modern structural geology, vol. 1: strain analysis, 307 p. (Academic Press, London).

RENARD, M. (1979): Aspect geochimique de la diagenese des carbonates.- Bull du B.R.G.M. (2) IV, 133-152.

RICKEN, W. (1985a): Epicontinental marl-limestone alternations: Event deposition and diagenetic bedding (Upper Jurassic, Southwest Germany).- in: BAYER, U. & SEILACHER, A. (eds.): Sedimentary and evolutionary cycles.- Lecture Notes in Earth Sciences 1, 127-162.

-- (1985b): Diagenetische Bankung: Zementbilanz, Geochemie und Fazies von Kalk-Mergel-Wechselfolgen.- Dissertation Univ. Tübingen, 242 p.

-- & HEMLEBEN, C. (1982): Origin of marl-limestone alternations (Oxford 2) in Southwest Germany.- in: EINSELE, G. & SEILACHER, A. (eds.): Cyclic and event stratification, 63-71 (Springer, Berlin).

RIEKE, H.H. & CHILINGARIAN, G.V. (1974): Compaction of argillaceous sediments.- Develop. Sedimentology 16, 424 p. (Elsevier, Amsterdam).

RITTENHOUSE, G. (1971a): Pore-space reduction by solution and cementation.- Am. Assoc. Petroleum Geol. Bull. 55, 80-91.

-- (1971b): Mechanical compaction of sands containing different percentages of ductile grains: a tneoretical approach.- Am. Assoc. Petroleum Geol. Bull. 55, 92-96.

ROBIN, P.Y.F. (1978): Pressure solution at grain to grain contacts.- Geochim. Cosmochim. Acta 42, 1382-1389.

ROLL, A. (1974): Langfristige Reduktion der Mächtigkeit von Sedimentgesteinen und ihre Auswirkung - eine Übersicht.- Geol. Jahrb. A14, 76 p.

ROGGENTHEN, W.M. & NAPOLEONE, G. (1977): Upper Cretaceous-Paleocene magnetic stratigraphy at Gubbio, Italy. IV Upper Maastrichtian-Paleocene magnetic stratigraphy.- Geol. Soc. Am. Bull. 88, 378-382.

SAITO, T. & BURCKLE, L.H. (1975, eds.): Late Neogen epoch boundaries.- Micropaleontology Press, New York).

SAYLES, F.L. & MANHEIM, F.T. (1975): Interstitial solutions and diagenesis in deeply buried marine sediments: results from the DSDP.- Geochim. Cosmochim. Acta 39, 103-127.

SCHLAGER, W. (1980): Mesozoic calciturbidites in deep sea drilling project hole 416A. Recognition of a drowned carbonate platform.- Init. Rep. DSDP 50, 733-749.

SCHLANGER, S.O. & DOUGLAS, R.G. (1974): Pelagic ooze-chalk-limestone transition and its implications for marine stratigraphy.- in: HSÜ, K.J. & JENKINS, C. (eds.): Pelagic sediments: on land and under the sea.- Intern. Assoc. Sedim. Spec. Publ. 1, 117-148.

SCHMIDT-KALER, H. (1962): Stratigraphische und tektonische Untersuchungen im Malm des nordöstlichen Ries-Rahmens.- Erlanger geol. Abh. 44, 51 p.

SCHMOKER, J.W. & HALLEY, R.B. (1982): Carbonate porosity versus depth: a predictable relation for South Florida.- Am. Assoc. Petroleum Geol. Bull. 66, 2561-2570.

SCHNEIDER, F.K. (1964): Erscheinungsbild und Entstehung der rhythmischen Bankung der altkretazischen Tongesteine Nordwestfalens und der Braunschweiger Bucht.- Fortschr. Geol. Rheinland u. Westfalen 7, 353-382.

SCHOLLE, P.A. (1977): Chalk diagenesis and its relation to petroleum exploration: Oil from chalks, a modern miracle?- Am. Assoc. Petroleum Geol. Bull. 61, 982-1009.

-- ; ARTHUR, M.A. & EKDALE, A.A. (1983): Pelagic sediments.- in: SCHOLLE, P.A.; BEBOUT, D.G. & MOORE, C.H. (eds.): Carbonate depositional environments.- Am. Assoc. Petroleum Geol. Mem. 33, 620-691.

SCHWARZACHER, W. & FISCHER, A.G. (1982): Limestone-shale bedding and perturbations of the earth's orbit.- in: EINSELE, G. & SEILACHER, A. (eds.): Cyclic and event stratification, 72-95 (Springer, Berlin).

SEIBOLD, E. (1952): Chemische Untersuchungen zur Bankung im unteren Malm Schwabens.- Neues Jahrb. Geol. Paläontol. Abh. 95, 337-370.

-- & SEIBOLD, I. (1953): Foraminiferenfauna und Kalkgehalt eines Profils im gebankten unteren Malm Schwabens.- Neues Jahrb. Geol. Paläontol. Abh. 98, 28-86.

-- & -- (1959): Kalkbankung und Foraminiferen.- Eclog. geol. Helv. 51.

SEILACHER, A.; ANDALIB, F.; DIETEL, G. & GOCHT, H. (1976): Preservational history of compressed Jurassic ammonites from southern Germany.- Neues Jahrb. Geol. Paläontol. Abh. 152, 307-356.

SHINN, E.A.; HALLEY, R.B.; HUDSON, J.H. & LIDZ, B.H. (1977): Limestone compaction - an enigma.- Geology 5, 21-24.

SIMPSON, J. (1985): Stylolite controlled layering in an homogenous limestone: pseudo-bedding produced by burial diagenesis.- Sedimentology 32, 495-505.

SPROVIERI, R. (1968): La serie Plio-Pleistocenica di Agrigento.- Giornale di Geologia 35, 295-301.

STEINEN, R.P. (1978): On the diagenesis of lime mud: scanning electron microscopic observations on subsurface material from Barbados, W.I.- Jour. Sed. Petrol. 48, 1139-1147.

SUJKOWSKI, Z.L. (1958): Diagenesis.- Am. Assoc. Petroleum Geol. Bull. 42, 2692-2717.

THIEDE, J.; VALLIER, T.L. et al. (1981): Site 463: Western Mid-Pacific Mountains.- Init. Repts. DSDP 62, 33-156.

THIERMANN, A. (1973): Erläuterungen zu Blatt 3710 Rheine.- Geol. Karte 1:2500 von Nordrhein-Westfalen, 174 p.

TRURNIT, P. & AMSTUTZ, G.C. (1979): Die Bedeutung des Rückstandes von Druck-Lösungsvorgängen für stratigraphische Abfolgen, Wechsellagerung und Lagerstättenbildung.- Geol. Rundschau 68, 1107-1124.

VINOPAL, R.J. & COOGAN, A.H. (1978): Effect of particle shape on the packing of carbonate sands and gravels.- Jour. Sed. Petrol. 48, 7-24.

VEIZER, J. (1977a): Diagenesis of pre-Quarternary carbonates as indicated by tracer studies.- Jour. Sed. Petrol. 47, 565-581.

-- (1977b): Geochemistry of lithographic limestones and dark marls from the Jurassic of southern Germany.- Neues Jahrb. Geol. Palaontol. Abh. 153, 129-146.

-- (1978): Simulation of limestone diagenesis - a model based on strontium depletion: Discussion.- Canadian Jour. Earth Sci. 15, 1683-1685.

-- & DEMOVIC, R. (1974): Strontium as a tool in facies analysis.- Jour. Sed. Petrol. 44, 93-115.

WALKER, R.G. (1984): Shelf and shallow marine sands.- in: WALKER, R.G. (ed.): Facies models (second edition), 141-170 (Geol. Assoc. Canada, Reprint Series 1).

WALTER, L.M. & MORSE, J.W. (1984): Reactive surface area of skeletal carbonates during dissolution: effect of grain size.- Jour. Sed. Petrol. 54, 1081-1090.

WALTHER, M. (1983): Diagenese gebankter Karbonate im Unterkarbon nordwest Irlands.- Thesis Univ. Göttingen, 76 p.

WANLESS, H.R. (1979): Limestone response to stress: pressure solution and dolomitization.- Jour. Sed. Petrol. 49, 437-462.

WEBER, R. (1951): Die Verbreitung von Pollen und Sporen in bituminösen Mergeln des Lias Alpha, Lias Epsilon und im Opalinuston Württembergischer Fundorte.- Thesis Univ. Stuttgart, 85 p.

WEDEPOHL, K.H. (1970): Geochemische Daten von sedimentären Karbonaten und Karbonatgesteinen in ihrem faziellen und petrographischen Aussagewert.- Verh. Geol. Bundesanstalt Wien 1970,4, 692-705.

-- (1979): Geochemische Aspekte der Diagenese von marinen Ton- und Karbonatsedimenten.- Geol. Rundschau 68(3), 833-847.

WEILER, H. (1957): Untersuchungen zur Frage der Kalk-Mergel-Sedimentation im Jura Schwabens.- Thesis Univ. Tübingen, 57 p.

WEPFER, E. (1926): Die Auslaugungsdiagenese, ihre Wirkung auf Gestein und Fossilinhalt.- Neues Jahrb. Mineral. Beilagen Bd. 54 B, 17-94.

WETZEL, A. (1981): Ökologische und stratigraphische Bedeutung biogener Gefüge in quartären Sedimenten am NW-afrikanischen Kontinentalrand.- "Meteor" Forsch.- Ergebnisse C34, 1-47.

WEYL, P.K. (1959): Pressure solution and the force of crystallization - a phenomenological theory.- Jour. Geophys. Research 64, 2001-2025.

WOLFE, M.J. (1968): Lithification of a carbonate mud: Senonian chalk in Northern Ireland.- Sedimentary Geology 2, 263-290.

ZANKL, H. (1969): Structural textural evidence of early lithification in fine-grained carbonate rocks.- Sedimentology 12, 241-256.

ZIEGLER, B. (1977): The "White" (Upper) Jurassic in southern Germany.- Stuttgarter Beiträge Naturkunde B26, 79 p.

SUBJECT INDEX